*Easy PC Astronomy* is the book for all those ⌐⌐⌐
calculations easily and accurately. A simple bu⌐⌐⌐⌐⌐⌐⌐⌐⌐⌐nguage called
*AstroScript* is provided on a disk with the book ⌐⌐ ιo use on any IBM PC-type
computer. Equipped with this the user can make complex calculations within
minutes, with no expert knowledge of astronomy, maths, or computer programming.
For example, a single-line program is all that is needed to see moving displays of
solar eclipses – whether that chronicled by the Chinese in 198 BC, or the millennium
eclipse to be seen in the UK in 1999. The *Sky graphics* facility displays a detailed
image of the sky as seen from any point on earth, at any time in the future or past
(showing the constellations, planets, and a host of other features), and updates
minute by minute if desired. For the expert, full details of the calculations (and
formulae) are included; and for the beginner, a comprehensive glossary of
astronomical terms is provided.

*Easy PC Astronomy* will be of immediate practical use to amateur astronomers (from
novice to advanced), students (in high school or university), science teachers, and
research astronomers.

*Easy PC Astronomy*

# *Easy PC Astronomy*

PETER DUFFETT-SMITH

*University of Cambridge*

Published by the Press Syndicate of the University of Cambridge
The Pitt Building, Trumpington Street, Cambridge CB2 1RP
40 West 20th Street, New York, NY 10011-4211, USA
10 Stamford Road, Oakleigh, Melbourne 3166, Australia

First published 1997

Printed in Great Britain at the University Press, Cambridge

*A catalogue record for this book is available from the British Library*

*Library of Congress cataloguing in publication data*

Duffett-Smith, Peter.
    Easy PC astronomy / Peter Duffett-Smith.
        p.    cm.
    Includes bibliographical references and index.
    ISBN 0 521 56052 7 (pbk.)
    1. Astronomy – Data processing.   2. AstroScript.   I. Title.
QB51.3.E43D84   1997
522'.85–dc20   96-2342CIP

ISBN 0 521 56052 7    paperback

*To all my colleagues at the Mullard Radio Astronomy Observatory*

# Contents

# *Preface*

This is a book for people who wish to make astronomical calculations quickly and accurately without first having to learn a lot of astronomy, mathematics, or computer programming. My aim has been to provide something that is of immediate use to both the unsophisticated and the expert alike, whether it be for amusement, instruction, or research.

I have already published two other books about astronomical calculating. The first of these, *Practical Astronomy with your Calculator*, provided simple methods which were not very precise. Even so, it would take some 20 minutes to find, say, the position of Venus pushing the buttons on your calculator, with always the possibility of error on the way. My second attempt, *Astronomy with your Personal Computer*, showed how to construct programs in BASIC using different subroutines to carry out each part of the astronomical calculation. Now the accuracy was much better, and of course the programs calculated very quickly, but it still took a long time to understand how to join the different parts of a program together. *Easy PC Astronomy* represents my third effort. The software is supplied on a disk with the book, and once you have loaded it onto your computer you should be able to make astronomical calculations within minutes using a simple high-level astronomical script language.

The astronomical script language is called *AstroScript* and is very easy to use, yet it provides results which are accurate enough for all but the most exacting astronomical purposes. For example, you can use AstroScript to display the total solar eclipse reported by the Chinese in the year 198 BC, or that which will be seen in Cornwall, England, on 11th August 1999. The AstroScript program you need to write to do this can be as simple as to contain just one command: *compute_eclipses(sun)*. All you need is this book, its disk, and an IBM compatible computer.

I am most grateful to the kind people who have helped me in the production of this book. In particular, I thank Dr Graham Woan who made me turn my rudimentary first version of AstroScript into something more useful, and Phil Whitehead who read the manuscript and tried the programs, making many improvements on the way. I am also grateful to Creidhe O'Sullivan, Mike Rushton, two anonymous referees, and colleagues at Cambridge for their suggestions and ideas. I would also like to thank, particularly, the copy editor at Cambridge University Press, Jo Clegg, who has made many improvements to the raw text. AstroScript is entirely original, but has leant heavily on the work of other authors whose methods and ideas I have adapted here and there in its construction. I am especially indebted to Jean Meeus and his excellent work *Astronomical Algorithms*, an example to any author of clarity, accuracy, and attention to detail. The list of stars was kindly provided courtesy of B. Yallop at the Royal Greenwich Observatory.

I have worked hard to make AstroScript as free from errors as possible, but I'm quite sure that bugs will come to light. I would therefore be grateful to hear from any reader who has found a problem, or who has suggestions to make for improvement. You can write to me at Downing College, Cambridge CB2 1DQ, UK, or send me e-mail via the Internet to PJDS@MRAO.CAM.AC.UK. Please supply your telephone number if writing by post. In view of the possibility of error, I have included the following disclaimer in AstroScript:

**AstroScript is provided as an aid to readers of *Easy PC Astronomy*. It is not free software, and may not be distributed to others nor modified without consent in writing from the author, Peter Duffett-Smith. Whilst every effort has been made to ensure the accuracy of AstroScript, no guarantee is given, neither implicit nor explicit, that it will function correctly in all circumstances, and so it should only be used as intended for recreational and other non-critical purposes.**

Peter Duffett-Smith
Downing College, Cambridge

# Introduction

There are lots of books available today about how to use your personal computer for making astronomical calculations. Most, like this one, come with a floppy disk to save you the tedium of typing everything in at the keyboard. Some include complete subroutines so that you can stitch together your own program with comparative ease. However, all of them suffer from the single disadvantage that they require a substantial degree of programming ability from you, the reader. Yet often you want to make some calculation quickly, perhaps to find the position of a planet or the phase of the Moon, and the task of writing a new program just gets in the way of the astronomy. This is true of my own book, *Astronomy with your Personal Computer*, which requires that you understand how the subroutines (written in BASIC) relate to each other, what variables you need to initialise, and how to manage the flags signalling the progress of the calculation, even before you can begin to write your program.

These drawbacks may make you turn to one of several glossy astronomy software packages that you can obtain for your pc. At first sight, you seem to be able to do all you would ever want with them, and as later versions come out, so you can do even more than before. Yet I find them disappointing in the end, not usually being able to supply what I actually need – at least not directly. I saw an example of this recently when trying to point a radio telescope towards one of the planets. The telescope control panel required me to supply the hour angle and the declination of the planet; my glossy package could show its position against a colourful background of stars, giving me pan, zoom, rotate, reverse, and real-time functions, but could supply only declination and right ascension as numbers. I therefore needed to find the local sidereal time (several key presses later) so that I could subtract the right ascension from it and obtain the hour angle. Of course, both the time and the right ascension were in hours, minutes, and seconds formats, making it just that much more difficult for me to get the right answer – and I was in a hurry!

*Easy PC Astronomy* represents my attempt to give you the ability to write your own astronomy programs with virtually no expertise. Rather than simply providing details of *how* to make calculations, I have devised a computer script language called 'AstroScript' which makes them for you. You want to find the sidereal time? Just write *find_time(sidereal)*. You need to calculate a solar eclipse? Just write *compute_eclipses(sun)*. Do you want the position of a minor planet next Thursday? Just write *find_position(elliptical)* if you have its osculating elliptical elements, or *find_position(parabolic)* if you have its parabolic elements. (See the *Glossary of astronomical terms* and later sections of the book for explanations of the meanings of these and other terms.) You need hardly any programming ability since the commands are straightforward and explicit, and you need only a rough knowledge of astronomy to put them in the correct order. Suppose, for example, that you wished to write a program to display the phases of the Moon in the current month. Your AstroScript program would be

    compute_moon_phases

and that's all. (Note the underline symbol used in many AstroScript commands.) AstroScript would ask you for all the details (such as the date) it needed to make the calculation. As embellishments, you might wish to add a title at the top of your screen and the facility to repeat the program for different dates, so you could then write

    display_title(The phases of the Moon)
    compute_moon_phases
    ask_for_repeat

You could use the program entering facility of AstroScript to enter these three lines of program, or an ordinary text editor to put these lines into a text file which you might call *phases.txt*. You would then run the program by typing *ascript phases.txt*. The computer would respond by displaying the following:

```
                        The phases of the Moon
                        ----------------------

    Please input an approximate date:
    Calendar date (d,m,y; BC neg) ............... ? 7/ 11/ 1994
    Time zone (h ahead of UT; W negative) ........ ? 0
    Daylight saving (h ahead of zone t) .......... ? 0

    (new page here)
                        ** The phases of the Moon **
                        ---------------------------

    Local circumstances of:

    New Moon ........................................    3-Nov-1994 at   13 37
    First quarter ...............................   10-Nov-1994 at    6 15
    Full Moon ...................................   18-Nov-1994 at    6 58
    Last quarter ................................   26-Nov-1994 at    7  5

    ** solar eclipse certain this month
    ** lunar eclipse certain this month

    Again (Y or N) ............................... ? n
```

(Things which the computer writes to the screen are in ordinary type, whilst your responses which you type at the keyboard are underlined.) The program first clears the screen and puts up the title *The phases of the Moon* centred and underlined at the top. It then asks you to supply a date, your time zone, and your daylight saving correction. In each case it offers you a default value which you can use simply by pressing the *enter* or *return* key, or you can put in your own value. It then clears the screen again, and writes the title ** *The phases of the Moon* ** centred and underlined at the top, followed by the local calendar dates and local civil times of the four phases of the Moon. It also determines whether an eclipse of the Sun or Moon is likely to occur. In this particular case, it correctly reports the total solar eclipse of November 3rd which was observed in Chile, and the partial lunar eclipse of November 18th. Finally, it asks whether you wish to repeat the program. You respond with *y* to run it again, say for another date or another time zone, or *n* if you wish to stop there. Note that AstroScript does not mind about upper and lower case letters here. You could respond with *N* or *n* with the same result.

The radio telescope pointing problem that I mentioned before would also have been easily solved by using

```
find_position(jupiter)
convert_coordinates(p,h)
display_title(The position of Jupiter)
display_coordinates(h)
wait_time_step(10)
repeat
```

This would have displayed the hour angle and declination of Jupiter on the screen, updating the coordinates every 10 seconds. Why not try this now for yourself?

Full details of AstroScript are given in the next chapter, and in *Appendices A*, *B*, and *C*. However, this book is not just a computer manual. I have taken care to describe as fully as possible the mathematical methods on which the calculations are based. If you want to write your own programs outside AstroScript, you may find this book a good starting place. But you need not absorb the mathematics in order to write programs in AstroScript. On the contrary, it has been my aim to provide something of immediate use to everyone, from novice to professional. Try it for yourself!

# Programming in AstroScript

AstroScript is an astronomical script computer language which has been designed to make astronomical calculations as easy as possible. If you are new to astronomy, you will be unfamiliar with many of the terms used. If you wish, you can read the later chapters of this book first in order to familiarise yourself with the subject before attempting to use AstroScript, but my strong advice is to try out AstroScript straight away. The best way to learn is by doing, and you can digest the complexities of astronomy more easily in small chunks. If you agree, then read on.

AstroScript supports about seventy commands, a quarter of which carry out specific astronomical calculations, the rest being concerned with displaying results or other 'housekeeping' duties. All the commands are linked together by means of an invisible data bank which holds the results of each calculation, making them available to subsequent commands. (More details of the contents of the data bank are provided in *Appendix C*.) Hence, if you run the command which calculates the position of the Sun, you can then run another which displays the result on the screen like this:

```
find_position(sun)
display_position(sun)
```

The second command looks for the position of the Sun in the data bank which has already been calculated by the first command. If you try to run *display_position(sun)* before *find_position(sun)*, you will get an error message.

This same data bank also stores all the details which you have to supply, such as your geographical position, time zone, and daylight saving correction. You need provide these only once, no matter how many different commands you use in your AstroScript program. Astronomical calculations are therefore quite straightforward, with the details kept strictly in the background and out of sight. You need to be concerned only with the general outline of the calculation, so that your program has its commands in the right order. You use a conventional text editor, such as DOS *Edit*, Windows *Notepad* or *Write*, or OS/2 *E*, to prepare a file of AstroScript commands in plain unformatted text, and then you run it (in DOS) using the *ascript* command. It is as easy as that!

Perhaps the power of AstroScript is best illustrated by means of examples. Suppose you wished to calculate the times of rising and setting of the Sun. You could use the following AstroScript program:

```
display_title(Sunrise and Sunset)
find_rise_set(sun)
```

```
display_rise_set(sun)
ask_for_repeat
```

The first command clears the screen and writes the title *Sunrise and Sunset* underlined and centred at the top. The next command actually does the calculation. Of course, it needs to know the date, geographical position etc., and since you haven't supplied these yet it asks you for them. Having made the calculation, the next command displays the results. The final command allows you to repeat the calculation if you so wish, say for another place on the Earth or for another date.

Here is a second example. This time, the task is to display the azimuth and altitude of the Moon in real time, using the computer's system clock to provide the local civil time, and recalculating the position every ten seconds. The following AstroScript program achieves this:

```
find_position(moon)
set_coordinates_type(topocentric)
convert_coordinates(m,a)
display_title(The position of the Moon)
display_coordinates(a)
wait_time_step(10)
repeat
```

The first command calculates the Moon's position, asking you to provide all the details for the calculation such as the date and the time, your position on the Earth etc. Next, the Moon's coordinates are converted from topocentric right ascension and declination into azimuth and altitude. (Do not worry at this stage if you are unfamiliar with the meanings of these terms. They are explained in later sections, and also in the *Glossary of astronomical terms*, but for now just accept them as examples in the use of AstroScript.) The two arguments contained within the brackets, 'm' and 'a', direct the command to make this particular conversion. The fourth command clears the screen and then displays the text contained within the brackets as a title, underlined and centred at the top. It is best to put this after the commands which actually carry out the computations since a slow computer may take several seconds to do so, and leave you looking at a blank screen for that time. The command *display_coordinates(a)* displays the results of the conversion (azimuth and altitude). Next, the *wait_time_step(10)* command suspends the program until the time on the computer's clock is 10 seconds later than the time used for the previous calculation. The local civil time is then set to this new value. The final command causes the entire program to be repeated from the top. This time, however, you will not be asked for any values since you have already provided them. The screen therefore shows the current altitude and azimuth of the Moon, updated every 10 seconds.

These two examples illustrate how each command in AstroScript carries out a host of different functions behind the scenes to achieve its results, worrying you with none of the details. However, if things go wrong such that it is impossible to carry out the command, AstroScript puts up an error message on the screen giving the reasons for the failure. For example, the following single-line script fails because you are trying to display a result before making the calculation:

display_position(jupiter)

When you run this script you get the error message *Jupiter's position invalid*, and the program then stops running.

AstroScript includes commands for carrying out almost all the calculations needed by the amateur astronomer, such as converting coordinates from one system to another, correcting for precession, geocentric parallax, and atmospheric refraction, converting between the Julian and Gregorian calendars, transforming sidereal times (local and Greenwich) into their corresponding solar times anywhere in the World and vice versa, finding circumstances of rising and setting for the Sun, Moon, major planets, and stars, calculating elliptical and parabolic orbits including those of the Sun, Moon, major and minor planets, asteroids, and comets, and calculating the circumstances of solar and lunar eclipses. Graphical displays are used where appropriate, for example in displaying a moving image of the progression of a solar eclipse. There is also a powerful facility for solving many *inverse* problems by iteration. You can use this to find the instant at which any member of the Solar System next reaches a particular value of a given coordinate, or when any two objects are in conjunction or opposition with each other. All AstroScript programs may be repeated under your control, or may be run automatically, with the results being continuously updated at predefined intervals. Results are displayed on the screen, but may also be duplicated to a disk file or to the printer, or to both. I have attempted to keep the syntax as simple as possible consistent with the range of tasks that must be addressed, incorporating only a limited amount of program flow control.

## *Installing AstroScript*

AstroScript is provided on the 3.5 inch diskette which accompanies this book. You **cannot** run the software directly from the diskette, but must first load it onto your hard disk (or other read–write medium). To do this you must use the program *install.exe* which is also on the diskette. The AstroScript files themselves are in coded form and called *easy_pc.nn*, where *nn* is a two-digit number (01, 02, 03 etc.). The installation program reads each file in turn, decodes it, renames it, and then deposits it in the target directory on the disk of your choice. Most machines have a 3.5 inch diskette drive called A:, and a hard disk called C:. We will suppose that this is the case in the example installation shown below. If your diskette drive is called B: and you wish to load AstroScript onto drive J:, just substitute the letters B for A and J for C.

```
*** AstroScript installation program ***
-----------------------------------------


In which drive is the AstroScript diskette (A, B, etc.) ? A
Destination drive for AstroScript (A, B etc.) .......... ? C
Directory in which to install AstroScript .............. ? C:\ASCRIPT
```

```
C:\ASCRIPT does not exist
Do you wish to create it (Y or N) ...................... ? Y

Your selections are:
Installing AstroScript from A:\
...................... to C:\ASCRIPT

Is this correct (Y or N) ............................. ? Y

Installing...
   AST_DESK.TXT..
   ORBITS.TXT..
   CALENDAR.TXT..
   GENCON.TXT..
   INVERSE.TXT..
   ECLIPSES.TXT..
   RISE_SET.TXT..
   PRECESS.TXT..
   EQ_TIME.TXT..
   REALTIME.TXT..
   TIMECON.TXT..
   ASCRIPT.EXE.............................
   SETUP.DAT..
   STARDATA.ASC..

Installation completed
Please remove the distribution diskette and keep safely
```

In this example, the text written by the computer is in normal type, whereas your responses are underlined. Hence to install your software you simply put the diskette into the 3.5 inch drive, select it (by entering *A:* or the appropriate drive letter), enter *install*, and then answer the questions posed on the screen as in the example above. (Don't forget to press the *enter* or *return* key after each command line you type at the keyboard.) The default values to the questions are shown in inverse type on the screen. You can edit these or use them directly – see the section below entitled *Input editing*. If you make a mistake, you can abort the installation by holding down the *cntrl* key and pressing *break*. If the program detects an error itself, it aborts the installation with an error message giving you the reason for the failure.

There is a small measure of copy protection incorporated in the software. This detects whether the physical location of the program *ascript.exe* has been changed, or whether the file itself has been modified. The first time you run *ascript* after installation you will be asked to register the program in your name. You need to supply your name in at least six characters, but not more than 16. 'Peter D-S', 'Sarah White', or 'Albert Einstein' are all acceptable names, but 'Jeff' (too short) and 'Alexander Fleming' (too long) are not. Be sure to write down your registration name (including upper and lower case letters, spaces, and any punctuation symbols) and keep it safely somewhere as you may need it again later, for example after running a defragmentation program. You will also be asked a question which you answer only if you have the book at hand. Such a question might be 'What is the last word of the section *Running AstroScript*?'. You need to find the word in the book, and enter it exactly as printed (including upper and lower case letters). The registration is completed by your answering this correctly, and you can then use the program as many times as you like without further interruption. However, if you copy the software to a new directory, or run a defragmentation program – in fact any operation which changes the physical location of *ascript.exe* on your disk – you will be asked to supply the registration name you used when you installed the software from the distribution diskette, and also be asked another question about the book. I hope very much that you don't find this procedure too intrusive. You should normally never meet it again after installation. Note that you can install the software as many times as you like from the original diskette, so if you forget your registration name, just start again from scratch.

## *Running AstroScript*

Your first task is to prepare a text file containing the AstroScript commands you wish to execute. You can do this either by using a text editor, or by entering the commands line by line using AstroScript's program entry facility (see below). The text file needs to be created in the directory which holds the *ascript.exe* file. This is normally on drive C: and called *ASCRIPT*. You can get to this by typing *C:* to select drive C:, followed by the DOS command cd \*ascript* to change directory to *ascript*. You can then run your editor to type in the program. For example, if you are using DOS version 5 or later you can use the *Edit* command. Decide what the name of the script file is to be, say *sunrise.txt*, and enter the line *edit sunrise.txt* at the DOS prompt, pressing the *return* or *enter* key at the end. Within a few moments you will see the edit program's screen, with the cursor positioned at the top of the editing space ready for you to type. Now you type in the AstroScript commands so that each is on a line by itself. Try typing in the four lines of the sunrise example given above, i.e.:

```
display_title(Sunrise and Sunset)
find_rise_set(sun)
display_rise_set(sun)
ask_for_repeat
```

When you have finished, exit from the editor by holding down the *alt* key and pressing *F*, then releasing the *alt* key and pressing *X*. You will be asked if you wish to save your new file, so answer *Y* (for yes).

If you are a Windows user, you can create your text file in a similar fashion using the Windows *Notepad* or *Write* commands (usually in the *Accessories* program group). When using *Write*, you must save your file with the *.txt* name extension in order that it is saved in simple text format. Remember to save your file in the *ASCRIPT* directory (or whichever directory holds the file *ascript.exe*). If you are using OS/2 you can use the editor *E*. Start a new OS/2 window, enter *E sunrise.txt*, and type in your commands as described above. Finish by saving the file (in the *ascript* directory) and exiting from *E* by double-clicking with the mouse pointer in the top left-hand corner of the OS/2 screen.

Having created your AstroScript program file, run it under DOS by entering

    ascript ⟨filename⟩

where ⟨*filename*⟩ is the name of your file (e.g. *ascript sunrise.txt* in this example). You need to be at the DOS prompt to do so, and in the directory containing both the *ascript.exe* file and your AstroScript file. Provided that ⟨*filename*⟩ exists, it is opened and interpreted from the top, line by line. Otherwise, if ⟨*filename*⟩ does not exist, either because you have not made it yet or because you misspelt it, the interpreter asks you if you wish to create it and enter your commands straight away. If you answer *Y* (for yes) you can use AstroScript's program entry facility to type in the commands line by line just as if you were using a text editor. There is a difference, however, because once you have pressed the *return* or *enter* key to finish a line, you cannot go back to it to correct any mistakes. (You can use an editor to do that later, or wait until the AstroScript interpreter finds your mistake and gives you the opportunity to correct it in the midst of running your program – see below.) When you have finished, type *Q* or *q* on a line by itself, and then the interpreter will begin running the commands line by line from the top. You can also specify whether your commands are to be saved in a permanent file to be used again later, or discarded at the end when the program finishes.

Here is an example. The file called *temp* does not already exist:

```
C:\ascript > ascript temp

This copy of AstroScript is registered to Peter D-S
Opening command file temp

   +++ temp not found

Do you wish to enter commands now (Y or N) ... ? Y
Save commands in temp (Y or N) ............... ? Y
Enter your commands; enter "Q" on an empty line to finish

display_title(Sunrise and Sunset)
find_rise_set(sun)
display_rise_set
ask_fer_repeat
q
```

As before, the computer's display is in normal type, and your responses are underlined. Having typed the last line (*q*), the AstroScript program runs lines one, two, and three as normal. However, when it gets to the fourth line, it stops and reports an error. (Notice that there is a mistake in the command; it ought to read *ask_for_repeat.*) You are then given the option of correcting the mistake and, further, saving the correction in file *temp*:

```
Error in command file at line 4
Command or its arguments not recognised

Command  "ASK_FER_REPEAT"
```

```
Correct the line (Y or N) .................... ? Y
>>  ask_for_repeat

Make the changes permanent ................... ? Y
Again (Y or N) ............................... ? n
```

Having replied *Y* to the question *Correct the line (Y or N)*, the AstroScript interpreter displays the faulty line and allows you to edit it. When you press the *enter* or *return* key to signify that you have finished editing, you are asked whether you wish to make the changes permanent. Answer *Y* to update the file (*temp* in this case) or *N* if you wish the file to remain as it was with the mistake in it. Whichever you answer, the AstroScript program then carries on with the line you have just edited.

If you type in the *ascript* command by itself, without giving a filename after it, the program assumes that you wish to run the default file called *acomm.txt*. If a file of this name exists, the interpreter opens it and runs the commands in it just as normal. Otherwise, you are given the option of entering your commands directly into *acomm.txt* line by line as above.


## *Input editing*

As your commands run, you may be asked to enter information via the keyboard. The program supplies a default value in every case which appears on the screen in inverse text after the prompt. (This is normally black characters on a light background, but you can change the text colour settings by editing the *setup.dat* file – see *Appendix B*, or using several *set* commands – see *Appendix A*.) You can select the default value simply by pressing the *return* or *enter* key. However, if you press one of the edit keys – *left arrow*, *right arrow*, *backspace*, *delete*, *insert*, or *cntrl-A* – you may then edit the default value, changing some or all of the characters as desired. The left and right arrow keys allow you to move the cursor backwards and forwards in the default entry. You will notice that new characters replace the old ones at the current cursor position; this is the *overtype mode*. You can also invoke the *insert mode* by which each character that you type is inserted into the line at the current cursor position with all the characters to the right of it being shifted right by one place. To do this, hold down the *control* key (cntrl) and press *A*, or press the *ins* key. An asterisk appears at the beginning of the entry, indicating that you are now in insert mode. Press *cntrl-A* or *ins* again to go back to overtype mode.

If the first key that you press is not an edit key, the default value disappears and your new entry appears in its place. However, there are two special cases. If you type *time* or *date* (and then press the *return* or *enter* key) your entry is replaced by a new default value equal to the time or the date respectively taken from the computer's system clock. This is especially useful if you wish to make a calculation for 'right now'.

## General comments

You can interrupt AstroScript programs at any time by holding down the *cntrl* key and pressing the *break* key once. This facility is always available in addition to the other program-control functions provided in AstroScript.

You will get best results with AstroScript by using a fast computer. Each calculation command carries out a large number of mathematical computations behind the scenes which may take some considerable time to complete on a slow machine. Although AstroScript will run on any IBM PC class computer and will get the answer you seek eventually, you may find yourself becoming frustrated by having to wait. I recommend strongly that you have a maths coprocessor (and a fast VGA screen if you wish to use the graphics commands). Whatever machine you use, you will need about 500 kB of free base memory in which to run AstroScript. This may require you to release memory by closing some transient programs loaded into the top end of your base memory by your *config.sys* and *autoexec.bat* files when you turn on your machine. Such programs are sometimes known as TSRs. You probably don't need most of them to run AstroScript, so 'comment them out' by inserting the letters 'rem' and then a space at the beginning of each line to be ignored. Alternatively, you may be able to load some of the TSRs into high memory. See your DOS manual for details, or run a memory management program such as *memmaker* to carry out the procedure automatically. If you modify your start-up files yourself, be sure to make backup copies first to which you can return if you get in a muddle. Having made modifications, you will need to restart your machine for them to take effect. Holding down the *cntrl* and *alt* keys, and then pressing the *del* key will usually accomplish this.

## AstroScript commands

AstroScript commands can be divided roughly into five classes: general housekeeping commands which set values or control the flow of the ascript program (e.g. *repeat*), commands whose sole purpose is to get new information from you via the keyboard (e.g. *ask_for_coordinates*), those which carry out specific calculations (e.g. *find_position*), commands which display information or results on the screen (e.g. *display_rise_set*), and commands which are so complex as to incorporate all of these classes at once (e.g. *compute_eclipses*). Many commands combine input and calculation functions together, so that if you order a particular calculation which requires information you have not yet supplied, the command asks you for it.

I have designed the structure of AstroScript to be as simple as possible. The script language is **not** a general purpose programming language like BASIC or Pascal, so you will not find it possible to write general programs with it. Many AstroScript commands carry out very high-level astronomical functions, saving you the tedium of coding your own programs and ensuring that no mistakes are made (other than mine of course!). If you are used to writing your own programs in a high-level language, you will find AstroScript quite restrictive, but I think it does have sufficient flexibility to do everything that the astronomer may need.

Unlike general purpose languages, you can use only **five** variables in your script program, two single character variables and three text strings. A single character variable can be set to represent any single character, such as 'Y', 'N', '£', or 'd'. The text strings, on the other hand, can be set to represent any sequence of characters lumped together, such as 'Yes', 'Moon', '13245.678', or '&*GHTY%'. The strings can have any length from 1 to 255 characters. They can be set, for example, to the names of objects whose positions are to be calculated, or can be used in *if* statements to control the flow of the script.

The character variables are c1 and c2, and you can assign any single character to these variables using one of a number of AstroScript commands. For example, if you wished to set c1 equal to the character 'h', you could write *set_character(c1,h)*, and to set c2 equal to a space you would write *set_character(c2, )*. The string variables are called s1, s2, and s3, and there are analogous *set* commands for them. Thus, to make s2 equal to 'The time of the conjunction is' you could write *set_string(s2, The time of the conjunction is)*. The character variables, c1 and c2, and the string variables, s1, s2, and s3, are all preset to the dot or point character '.' when an AstroScript program begins. You do not have access to any other program variables except via the high-level input and output commands.

AstroScript commands often carry *arguments* which modify their actions. These are represented within brackets after the command itself. The commands *set_character* and *set_string* mentioned above are two such commands. In these particular cases, the strings of characters within the brackets have two parts, separated by a comma. Each represents one argument. Thus the command *set_character(c1,%)* carries the arguments 'c1' and '%', often referred to as argument 1 and argument 2. AstroScript commands may carry zero, one, two, or three arguments, separated by commas (but see the command *set_delimiter* in *Appendix A* and *delimiter* in *Appendix B*, both of which allow you to change the character separating the arguments from a comma to any other character). If the command carries zero arguments, such as *ask_for_repeat*, the brackets may be left out if you wish. If they are included, they should be empty. Thus *ask_for_repeat* and *ask_for_repeat()* are both legal and are equivalent to one another. Commands and their arguments may be entered in upper case, lower case, or any combination. The commands *convert_coordinates(m,a)* and *ConVert_Coordinates(m,A)* are equally valid.

The script file may contain any number of blank lines between lines of valid AstroScript commands. You may insert a comment in the file by beginning the line with the character '{' (and, optionally, ending the line with the character '}'). All lines beginning with '{' are ignored by the interpreter. There can be at most just one AstroScript command per line. There may be any number of spaces before each command, so you may indent the commands if you wish to make the program more readable. Spaces may also be inserted between the command and the opening bracket containing its arguments. You can also use the rest of the line after the closing bracket as you wish, since any characters after it are ignored. Hence the following are valid AstroScript lines:

>{This line is a comment}
>ask_for_string (What is the planet's name,s1) {This is also a comment}

whereas this line is not valid because the command has no argument:

>compute_moon_phases {This is an illegal comment}

Note however that you can give a *null argument* to any of the commands which do not normally take an argument by including an empty matched pair of brackets after it. Hence if you really must include a comment after the compute_moon_phases command you could write:

>compute_moon_phases() {This comment is OK}

The best way to learn AstroScript is by using it. I have provided many example AstroScript programs within the text, and some of these are also on the disk so that you can run them straight away. Try them out for yourself. Examine them to see how they work, and modify them as you wish. Remember that you can always reload the software from your distribution diskette if you get into a complete muddle, so don't be afraid to experiment. Every AstroScript command is described in detail in *Appendix A*, and you will need to refer to this in order to find out exactly what each command does. But for now, here is a list of all the AstroScript commands:


GENERAL HOUSEKEEPING COMMANDS:

*copy_to_printer( ⟨string ⟩)*
*copy_to_file( ⟨string ⟩)*

*if ( ⟨char variable or string variable ⟩, ⟨condition ⟩, ⟨char or string ⟩)*
*end_if*

*repeat*
*ask_for_repeat*
*repeat_for_count( ⟨string ⟩)*

*wait_time_step( ⟨string ⟩)*
*pause*
*pause( ⟨string ⟩)*
*suspend( ⟨string ⟩)*

*set_string( ⟨string variable ⟩, ⟨string ⟩)*
*set_character( ⟨char variable ⟩, ⟨char ⟩)*

*set_dynamic_offset(⟨string⟩)*

*set_delimiter(⟨char⟩)*
*set_decimal_places(⟨string⟩)*
*set_date_format(⟨string⟩)*
*set_normal_text(⟨string⟩)*
*set_inverse_text(⟨string⟩)*
*set_normal_background(⟨string⟩)*
*set_inverse_background(⟨string⟩)*
*set_aspect_ratio(⟨string⟩)*

*set_maximum_step(⟨string⟩)*
*set_number_of_iterations(⟨string⟩)*
*set_tolerance(⟨string⟩)*
*set_output_file_name(⟨string⟩)*
*set_longitude(⟨string⟩)*
*set_latitude(⟨string⟩)*
*set_height(⟨string⟩)*
*set_time_zone(⟨string⟩)*
*set_daylight_saving(⟨string⟩)*
*set_vertical_shift(⟨string⟩)*
*set_local_civil_time(⟨string⟩)*
*set_calendar_date(⟨string⟩)*

*increment_normal_text(⟨string⟩)*
*increment_inverse_text(⟨string⟩)*
*increment_normal_background(⟨string⟩)*
*increment_inverse_background(⟨string⟩)*

*increment_longitude(⟨string⟩)*
*increment_latitude(⟨string⟩)*
*increment_height(⟨string⟩)*
*increment_time_zone(⟨string⟩)*
*increment_local_civil_time(⟨string⟩)*
*increment_calendar_date(⟨string⟩)*

### INPUT COMMANDS:

*ask_for_character(⟨prompt⟩, ⟨accept⟩, ⟨char variable⟩)*
*ask_for_string(⟨prompt⟩, ⟨string variable⟩)*
*ask_for_coordinates(⟨char⟩)*
*clear_flags(⟨string⟩)*
*edit_star_data*

### CALCULATION COMMANDS:

*convert_coordinates(⟨char⟩, ⟨char⟩)*
*set_coordinates_type(⟨string⟩)*

*correct_for_precession*
*correct_for_parallax(⟨string⟩)*
*correct_for_refraction(⟨string⟩)*

*find_time(⟨string⟩)*
*find_date(⟨string⟩)*

*find_rise_set(⟨string⟩)*
*find_position(⟨string⟩)*
*find_when(⟨string⟩, ⟨string⟩, ⟨string⟩)*
*set_messages(⟨string⟩)*

### DISPLAY COMMANDS:

*new_line*
*write_string(⟨string⟩)*
*display_title(⟨string⟩)*

*display_date(⟨string⟩)*
*display_time(⟨string⟩)*
*display_time_difference(⟨string⟩, ⟨string⟩)*

*display_coordinates(⟨char⟩)*
*display_data(⟨string⟩, ⟨string⟩)*

*display_rise_set(⟨string⟩)*
*display_position(⟨string⟩)*

### COMPLEX COMMANDS:

*compute_nutation*
*compute_obliquity(⟨string⟩)*
*compute_moon_phases*
*compute_ephemeris(⟨string⟩)*
*compute_eclipses(⟨string⟩)*
*compute_sky*

# *Time*

We all know what we mean by *time*. It is that ever-flowing, forward-moving entity which orders the events in our lives. It is easy for us to say that one event is *later* than another, or that this moment occurred *before* that one. We are able to discern the direction of time very easily, to chop it up into little bits, each joined to the next, and each one strictly in serial order. Yet this time with which we are all so familiar is nevertheless deeply mysterious. We cannot see it, we cannot hold it in our hands, we cannot (or so it seems at first) influence its steady passage at all. But we can *measure* its passing very accurately; indeed it is the quantity that can be measured most accurately of all, more so than, for example, weight or temperature. The best Earth-based clocks are atomic clocks which use the oscillations associated with certain atoms to divide each second into thousands of millionths. Such is their precision that the error at the end of fifteen thousand million years (roughly the age of the Universe) may be no more than an hour or so. Many such clocks, kept in carefully controlled conditions at observatories around the world, now form the basis of civil time-keeping on the Earth, though even they might be superseded in years to come by clocks using pulsars as controlling mechanisms.

When we measure time, we make the basic assumption that the flow of time is uniform. If we can find a process which repeats itself exactly, like the swing of a pendulum, the rotation of the Earth, or the oscillation of an atom, then we can divide up the passage of time into lots of sequential intervals, each exactly the same length as the others. Our natural inclination is to assume that time is absolute, that is two perfectly accurate clocks in different places would remain in perfect synchronism for ever. But Einstein's theory of relativity, developed at the beginning of the twentieth century, demonstrated that this assumption is false. The rate at which a perfect clock keeps time relative to another depends upon both its relative speed, and its gravitational potential. This deduction arises from the observation that the speed of light in a vacuum appears to be the same no matter how the observer moves with respect to its source. For example, if the observer were to move in a vacuum at nine-tenths the speed of light towards a lamp, he would nevertheless measure the speed of the light coming from the lamp towards him to be exactly the same as if he were standing still. Einstein showed that, as a consequence, moving clocks appear to run slow. This was no trick, nor optical illusion. The same deduction applied to *all* clocks, whether mechanical, atomic, or biological.

The theory of relativity has virtually no consequence for our everyday lives. Even if you spent all your life travelling at 500 miles per hour in a jet going round and round the Earth, you would age by only 1 part in a million million more slowly than your brother or sister, and hence would live longer by only about one-fifth of a second. However, astronomers often wish to make measurements right at the limit of the available accuracy, so have to be concerned about the reference frame in which to work. Hence several different time frames have been defined, including one appropriate to the centre of the Earth, and another appropriate to the centre of our Solar System. For the most part, we shall not be concerned by such tiny differences and will be able to assume that identical clocks anywhere in the Solar System all advance at the same rate.

## Civil time

Our everyday lives are regulated by a locally agreed time scale called *local civil time (LCT)*. When we say that the time is ten thirty am, or fifteen twenty-five, we are usually making a statement about LCT. It is the time given out on the radio and television, or by the speaking clock on the telephone, and is usually the time to which you set your watches and clocks. LCT depends among other things on the time of year and your position on the Earth. The reference line for civil time-keeping is (for historical reasons) the line of longitude 0° which joins the north and south poles of the Earth and passes through a point in Greenwich, England. This line is called the *Greenwich meridian*. The Earth's surface is divided into (roughly) north–south strips or *time zones* in which the local civil time is the same and offset by a fixed number of hours from the time at Greenwich (see Figure 1). The offset, or *zone correction*, is usually chosen such that the Sun reaches its highest point in the sky (when it crosses the *local meridian*) in the middle of the zone at roughly 12 o'clock zone time. The zone time is often known as *standard time*, and is usually the value of LCT. However, many countries adopt the convention of advancing the local civil time by one hour during the summer months in order to make the daylight hours fit more conveniently into the civil day. The local civil time is then often known as *daylight saving time*, or *summer time*.

If the zone time at Greenwich is ZTG, the time zone correction is TZC, and the daylight saving correction is DSC, then the local civil time is given by

$$LCT = ZTG + TZC + DSC$$

Time zones east of Greenwich have positive values of TZC, and time zones west have negative values. AstroScript takes care of these details for you. When it needs to know how to convert the local civil time, it asks you for both your time zone correction and daylight saving correction.

The Greenwich zone time is often called *Greenwich mean time (GMT)*, an ambiguous term which had a different meaning before 1925 from its present-day definition. GMT is now set equal to the *coordinated universal time (UTC)*, which is a time derived directly from the network of atomic standards in observatories around the world, and is often referred to simply as universal time, UT, although there are strictly small

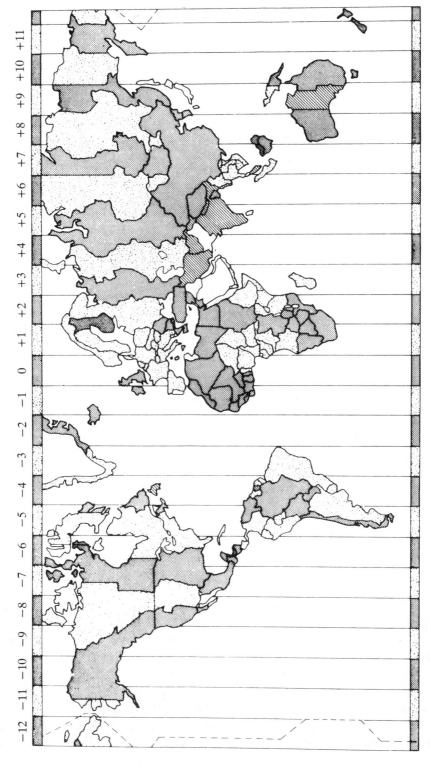

Figure 1. International time zones. This small-scale map can show only the general distribution of time zones around the world. If you are unsure of your own zone correction, you can check it by tuning your short-wave radio to the BBC World Service and comparing your watch with the time pips broadcast every hour from London.

differences between various flavours of universal time. The analysis of those atomic standards gives *international atomic time (TAI)*, which is as smooth and regular a time scale as we currently know how to measure. However, our daily lives have more to do with the position of the Sun in the sky and the season of the year than with processes going on inside atoms, and these are things which depend on the rotation of the Earth. Indeed, we used to measure time with respect to the rotation of the Earth, each rotation being called one day and divided into 24 hours of 60 minutes of 60 seconds. We now know that the Earth's spin is decreasing gradually with unpredictable irregularities in the slow-down, so that if we were to use TAI without correction, civil time would gradually get out of step with the daylight hours. UTC is therefore derived from TAI by adding a fixed number of seconds to it such that UTC is within one second of what we would have measured using the Earth as the time-keeper. This process requires the addition of an extra leap-second every year or two, usually at the end of June or December. On July 1st 1993, UTC was 28 seconds behind TAI. Thus UTC is an atomic time (and hence as smooth as we can measure) but with jumps in it to keep it in line with the irregular rotation of our planet.

## Dynamical time

In all our theories about the movement of celestial bodies such as the Moon and the planets, and indeed about any natural processes, we assume that there exists a perfectly uniform and smooth time which, if only we could measure it, would fix the position of the body exactly. We call this *dynamical time*. For example, we may have a theory about the motion of a pendulum bob swinging back and forth on a perfect grandfather clock. In principle, we could use this theory to predict exactly the position of the bob, say, 1000 years in the future, but in so doing we have to make the assumption that the time flows regularly. Yet we do not know how to measure anything so perfectly. TAI is the closest approximation we have to a perfectly smooth scale at present, so we measure dynamical time using it. The appropriate dynamical time for the apparent geocentric positions (measured with respect to the centre of the Earth) of bodies in the Solar System is called *terrestrial dynamical time (TDT)*, and is set to TAI + 32.184 seconds. The offset of 32.184 s was added to TAI in order that TDT carried on smoothly from *ephemeris time (ET)*, which was its equivalent before 1984. (Ephemeris time was deduced from the motion of the Moon in the days before atomic clocks became so accurate.) Strictly, then, if we want to calculate the position of a body like the Moon at a given local civil time, we should first convert to universal time, then make the correction from universal time to dynamical time, and then use the dynamical time as the basis of our calculations.

The difference between universal and dynamical times reached very large values in the past (e.g. about 4 hours in 500 BC), and may again reach large values in the future. However, we cannot estimate this difference with precision far ahead of the present date because the irregularities in the Earth's spin seem to be more or less random.

People have deduced the difference, $\Delta T$ = dynamical time − universal time, from historical records for dates in the past, and AstroScript uses a series of approximate expressions to estimate this difference for all dates prior to 1988 as follows.

For years from 1800 to 1988 inclusive:

$$t = (\text{julian\_date} - 2\,415\,020.0)/36\,525.0$$

$$\Delta T = -0.000\,014 + 0.001\,148t + 0.003\,357t^2 - 0.012\,462t^3 - 0.022\,542t^4$$
$$+ 0.062\,971t^5 + 0.079\,441t^6 - 0.146\,960t^7 - 0.149\,279t^8$$
$$+ 0.161\,416t^9 + 0.145\,932t^{10} - 0.067\,471t^{11} - 0.058\,091t^{12}.$$

This approximation was devised by Schmadel & Zech, has a maximum error of 1.9 s, and gives the value of $\Delta T$ in days. The symbol julian\_date here refers to the number of Julian days which have elapsed since Greenwich mean noon on January 1st, 4713 BC (see later in this chapter).

For the years from 1700 to 1799 inclusive:

$$\Delta T = 9 + (\text{julian\_date} - 2\,341\,972.5)/5217.86,$$

and from 1600 to 1699 inclusive:

$$\Delta T = 9 + (\text{julian\_date} - 2\,341\,972.5)^2/7\,424\,421.0.$$

I devised these two expressions myself. They are accurate to about 10 seconds over most of the range, and $\Delta T$ is expressed in seconds.

For years from AD 948 to AD 1600 inclusive:

$$t = (\text{julian\_date} - 2\,451\,545.0)/36\,525.0$$

$$\Delta T = 50.6 + 67.5t + 22.5t^2,$$

and for all dates before AD 948:

$$t = (\text{julian\_date} - 2\,451\,545.0)/36\,525.0$$

$$\Delta T = 2715.6 + 573.36t + 46.5t^2.$$

These two expressions are due to Stephenson & Houlden, and give $\Delta T$ in seconds. Their precisions become increasingly uncertain the further into the past they are used, but Stephenson & Houlden estimate that errors are likely to be no more than about 80 s at AD 948, 7 minutes at 390 BC, and about 15 minutes at 1500 BC.

The value of $\Delta T$ is set to zero for all dates from 1989 onwards, but you can set it to any other value using the *set_dynamic_offset* command (see *Appendix A*). Indeed, you can use this command for any calculation in the future or the past. Thus if you **know** what the difference between the dynamic and universal times was for a particular calculation, you can use the command to set your own value and by-pass AstroScript's built-in estimators. If you wish to turn off the dynamical time adjustment, simply write *set_dynamic_offset(0)* at the top of your AstroScript program.

More details of the complexities of time scales can be found in the *Explanatory Supplement to the Astronomical Almanac* (see *Bibliography*).

## Sidereal time

The previous section described how the local civil time is regulated by the apparent motion of the Sun around the Earth. We can imagine a fictitious body, called the *mean Sun*, which moves uniformly along the celestial equator (the line of intersection of the Earth's equator with the celestial sphere – see the next chapter) throughout the year, and whose passage across the Greenwich meridian each day corresponds closely with twelve o'clock UTC. Astronomers are more often concerned with the positions of stars and other bodies in the sky than they are with the Sun, and it is convenient to define another time scale called *sidereal time* for this purpose. During the course of one year, the Earth turns on its axis about 366.25 times, and the distant 'fixed' stars therefore appear to circumscribe the celestial poles this many times. But the Earth also makes one complete orbit of the Sun in the same year, and hence the Sun appears to make one less revolution of the Earth. Hence there are 365.25 solar days of successive passages of the (mean) Sun across the Greenwich meridian, and 366.25 star or *sidereal* days of successive passages of a reference star during the course of the same year. Sidereal time is measured in sidereal days, hours, minutes, and seconds, similar to solar time, but of course the sidereal quantities are shorter by about the factor 365.25/366.25, i.e. 1 second of solar time is equal to 1.002 738 seconds of sidereal time.

The reference 'star' chosen for sidereal time is not actually a star at all, but is a fixed direction called the *first point of Aries*, or *vernal equinox*, and the local sidereal time is defined to be its *hour angle*. (These terms are explained in the next chapter.) Solar time and sidereal times at Greenwich agree at one instant every year (around September 22nd) when the (mean) Sun is 12 hours distant from the vernal equinox. Thereafter, the difference between them grows by about 4 minutes per day in the sense that sidereal time runs faster than solar time.

The conversion between Greenwich sidereal time and universal time is governed by the following formula which was adopted by the International Astronomical Union in 1982. First find the number of Julian centuries, $T$, of 36 525 days since the epoch J2000, i.e. since the moment of midday at Greenwich on January first in the year 2000. This is best achieved by using *Julian day numbers* (see below). If *JD* is the Julian day number at 0 h UT on the date for which we wish to find the sidereal time, then

$$T = \frac{JD - 2\,451\,545.0}{36525},$$

and the mean sidereal time at Greenwich, *GST*, is given by

$$GST = 6^h41^m50^s.548\,41 + 8\,640\,184^s.812\,866\,T + 0^s.093\,104\,T^2$$
$$- 0^s.000\,0062\,T^3.$$

AstroScript uses this expression internally in its procedures for converting between solar and sidereal times.

Your *local sidereal time (LST)* is your local hour angle of the vernal equinox, and the *Greenwich sidereal time (GST)* is this hour angle as measured at Greenwich. These two are different in hours by your longitude in degrees divided by 15, i.e.

$$LST = GST + (longitude/15),$$

where longitudes **east** are **positive** and west are negative. (This is because the equator is divided into 360 degrees of geographical longitude and takes 24 sidereal hours to turn once on itself: $360/24 = 15$.)

The AstroScript command for converting between solar and sidereal times is *find_time($\langle string \rangle$)*, where $\langle string \rangle$ is 'solar' to convert from sidereal to solar, and 'sidereal' to convert from solar to sidereal. There is also the display command *display_time($\langle string \rangle$)* to display the result of the conversion. Thus to convert from solar to sidereal time, and then display the result you could write

```
find_time(sidereal)
display_time(sidereal)
```

However, you will probably want to use a more flexible program than that which converts in either direction under your control. The disk which comes with this book carries an example AstroScript program called *timecon.txt* which is as follows:

```
{Display a title and clear the screen}
display_title(Solar and sidereal Times)

{Which way round do we wish to convert?}
ask_for_character(Convert from solar to sidereal [Y or N],yn,c1)
if (c1=Y)
   set_string(s1,sidereal)
   set_string(s2,solar)
end_if
if (c1!Y)
   set_character(c1,Y)
   ask_for_character(....... from sidereal to solar [Y or N],yn,c1)
   if (c1=Y)
      set_string(s1,solar)
      set_string(s2,sidereal)
   end_if
end_if
```

```
{If we made a selection, carry out the conversion and display the result}
  if (c1=Y)
    new_line
    find_time(s1)
    new_line
    display_time(s2)
    display_time(s1)
  end_if

{Another go?}
  new_line
  ask_for_repeat
```

Here the value of c1 controls the direction of the conversion. If the answer to the first question is 'Y', then s1 is set to 'sidereal' and the conversion made from solar to sidereal time (*find_time(s1)*). If the answer is 'N' (i.e. c1!Y, c1 does not equal 'Y'), the second question is asked. If you now reply 'Y', s1 is set to 'solar' and the conversion is made from sidereal to solar. If you reply 'N' again, no conversion is made at all. The string s2 is set to the opposite of s1, i.e. if s1 is 'solar' then s2 is 'sidereal' and vice versa. This allows both types of time to be displayed together by the lines *display_time(s2)* and *display_time(s1)*.

There is a problem in converting from sidereal time to solar time. The length of the sidereal day of 24 sidereal hours is about 4 minutes shorter than the calendar day of 24 solar hours. Hence, on a given *calendar* date, a small range of sidereal times occurs twice (see Figure 2). Unless you know the Greenwich sidereal day number (unlikely) you cannot resolve the ambiguity. AstroScript therefore displays both sidereal times whenever the ambiguity arises, and leaves you to choose between them. For example, try converting from local civil time $0^h\ 2^m\ 0^s$, on date 8th September 1994, in time zone 0, with daylight saving 0, and longitude $0°\ 0'\ 0''.0$. You will find that the local sidereal time is $23^h\ 9^m\ 18^s.63$. Now convert back the other way, and you will be presented with the two possible local civil times, $0^h\ 2^m\ 0^s.0$ and $23^h\ 58^m\ 4^s.09$. Both are correct in that they both correspond to the given local sidereal time on the given calendar date.

## *Equation of the equinoxes*

The AstroScript algorithms calculate the *mean* sidereal time, that is the hour angle of the intersection of the ecliptic with the mean equator. But the Earth exhibits a slight wobbling motion called nutation which causes the instantaneous apparent direction of this line of intersection to be different by the small amount $\delta t$ seconds of time, where

$$\delta t = \Delta\psi \cos\varepsilon/15,$$

$\Delta\psi$ is the *nutation in longitude* in arcseconds, and $\varepsilon$ is the *obliquity of the ecliptic* (see next chapter). These two quantities may be calculated using the AstroScript commands

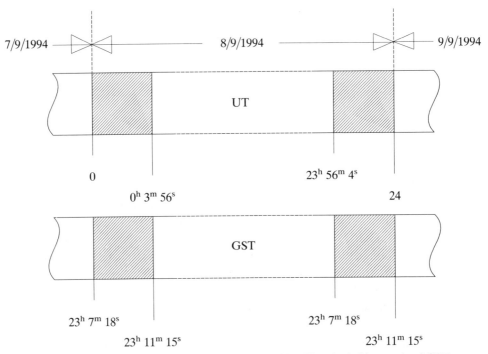

Figure 2. UT and GST for September 8th 1994. The shaded intervals of GST occur twice on the same day.

*compute_nutation* and *compute_obliquity*. The apparent sidereal time is then found by adding δt to the mean sidereal time, a correction known as the *equation of the equinoxes*. (You can use the AstroScript command *display_time_difference* to add two times together or subtract them from each other – see *Appendix A*.)

## *Equation of time*

The Earth's orbit about the Sun is not quite circular, but is an ellipse with an eccentricity of about 0.016 713. This causes the Earth–Sun distance to vary by about 2 million miles or so during the course of the year. Furthermore, the Earth's north–south spinning axis is tilted away from the perpendicular direction to the ecliptic plane by about 23°.5. These two factors cause the apparent motion of the Sun around the Earth to vary in speed during the year. We have already mentioned the concept of the *mean Sun*, which is a fictitious body moving in a circular orbit along the equator whose position coincides exactly with the true Sun once per year. This mean Sun moves at exactly the same rate throughout the year, and is therefore sometimes slightly ahead of the true Sun position, and sometimes slightly behind. Since the civil time is closely allied with the mean Sun, it means that noon as judged from the passage of the true Sun across the local meridian will usually not occur at 12.00 civil time. The difference between the mean Sun time (MST) and the real Sun time (RST) is called the *equation of time*, *E*, i.e.

$$E = \text{MST} - \text{RST}.$$

RST is what you would measure with your sundial, and MST is what your watch would read (due allowances having been made for daylight saving correction).

AstroScript provides commands for calculating the position of the Sun, and hence for displaying the equation of time. The key command is *find_when(sun,hour angle,0)*. This calculates the moment closest to a given starting point when the Sun's hour angle is zero, i.e. when it crosses the meridian. The difference between this time and 12.00 (the time when the mean Sun crosses the meridian) is the equation of time, and it may be displayed using the command *display_time_difference*. The disk which comes with the book carries a program to do all this for you called *eq_time.txt* as follows:

```
{AstroScript program to calculate the equation of time = MST - RST}

{Set the default values for the various quantities}
set_time_zone(0)
set_daylight_saving(0)
set_longitude(0,0,0)
set_latitude(0,0,0)
set_height(0)
set_decimal_places(0)
set_local_civil_time(12,0,0)

{Clear the screen and display the title in light grey on black}
set_normal_text(7)
display_title(The equation of time)

{Find the moment when the Sun crosses the meridian}
find_when(sun,hour angle,0)

{Display the result in red}
new_line
set_normal_text(4)
write_string(The equation of time [h,m,s] is ............. )
display_time_difference(12.0,local civil time)
set_normal_text(7)
new_line
new_line

clear_flags(all)
ask_for_repeat
```

You have to specify the date on which to find the equation of time. You are also asked for a time. Use the default value of 12.00 for this as the real Sun's passage across the meridian is never more than about 20 minutes either side of twelve o'clock. Note the use of the *clear_flags(all)* command before the *ask_for_repeat* command to make sure that the program asks you for a new date each time through. When I ran this script on 26th March, 1995, I obtained the following:

```
                              The equation of time
                              --------------------

Please input a starting point:
Calendar date (d,m,y; BC neg) ................ ? 26/ 3/ 1995
Local civil time (h,m,s) ..................... ? 12 0 0


The calculation converges on the epoch when:
The local calendar date is ...................     Sunday 26-Mar-1995
The universal time is (h,m,s) ................      12  5 50
The local civil time is (h,m,s) ..............      12  5 50
(No estimate available for ephemeris/dynamic time)

The equation of time [h,m,s] is .............. -    0  5 50

Again (Y or N) ............................... ? n
```

Note that input requiring several values, such as the date and time above, can use commas, slashes, or spaces to separate the numbers, e.g. 26/3/1995 or 26 3 1995 or 26,3,1995.


## *Calendars*

The local civil time is, by itself, not sufficient to define any instant uniquely. Obviously a given time repeats itself every day of twenty-four hours, so we need an additional system of reckoning days. There are many such systems, but the one in general use today is known as the Gregorian calendar, since it was devised through a modification of another system by Pope Gregory in 1582. This new calendar was gradually adopted by countries in Europe during the following centuries (United Kingdom in 1752, Turkey in 1927). Today it is used virtually world-wide for civilian time-keeping, though there are also many other religious calendars which are kept in various countries.

In the Gregorian calendar, the year has either 365 or 366 days in it, which are grouped into 12 months of 28, 29, 30, or 31 days each as follows:

|           | Normal year | Leap year |           | Normal year | Leap year |
|-----------|-------------|-----------|-----------|-------------|-----------|
| January   | 31          | 31        | July      | 31          | 31        |
| February  | 28          | 29        | August    | 31          | 31        |
| March     | 31          | 31        | September | 30          | 30        |
| April     | 30          | 30        | October   | 31          | 31        |
| May       | 31          | 31        | November  | 30          | 30        |
| June      | 30          | 30        | December  | 31          | 31        |

The month names are those devised by Julius Caesar in his calendar of about AD 8. A given month always has the same number of days in it every year, except for February, which has an extra day added to it every 4 years or so. This complication is necessary to accommodate the fact that the calendar needs to have an integer number of days in each calendar year, but the Earth actually takes about 365.2422 days (the *tropical year*: see the *Glossary of astronomical terms* for more information about the length of the year) to complete one orbit of the Sun. Pope Gregory's rule for adding the extra day is as follows:

> The year is deemed to be a leap year in which the month of February has 29 days (rather than its more-usual 28 days) if the year number is divisible by 4, except if the year number ends in two zeroes, in which case the year is a leap year only if it is also divisible by 400.

Thus the years 1992, 1836, and 2000 are all leap years, whereas 1991, 1837, and 1900 are not. Four hundred civil years therefore contain $(400 \times 365) + 100 - 3 = 146\,097$ days, and the average length of the civil year is $146\,097/400 = 365.2425$ days, a fair approximation to the length of the tropical year.

Julius Caesar's calendar was in use in Europe before 1582, and was similar to the Gregorian calendar except that *all* years divisible by four were leap years. Pope Gregory simply introduced his new rule and decreed that the date following 4th October 1582 was to be 15th October 1582 in order to realign the civil date with the seasons of the year. The dates 5th October to 14th October 1582 inclusive never existed. (This apparently caused some commotion amongst the people who thought they were being robbed of time.) The starting point, year AD 1 (*anno domini*), was selected to be the year in which Jesus Christ was born. Years were then counted sequentially. We therefore understand the date 19th May AD 1977 to mean the 19th day in the month of May of the 1977th year after the adopted instant of the birth of Christ, counting the birth year as 1. There is some confusion about the counting of the years preceding AD 1, since historians miss out year zero, i.e. the year immediately before AD 1 is 1 BC (*before Christ*). Astronomers and computers count logically, however, and consider that the year before AD 1 is 0, and before that is −1 etc. AstroScript takes care of this for you. It adopts the historians' convention of BC years, and recognises that there is no year zero. You simply enter any BC year with a minus sign in front of it. Thus the year 413 BC would be entered as −413, and AD 413 as +413 etc. Whenever AstroScript accepts a negative year, it adds one internally to convert from the historians' reckoning to the sequence of integers. Similarly, whenever a zero or negative year number is to be displayed, AstroScript subtracts one from it, removes the minus sign, and appends the letters BC after it.

## *Julian day number*

For calculation purposes, it is best to have a system of reckoning days from some starting point in which each day number follows logically on from the previous day number without any complications of month name, year number, or papal decree.

Astronomers have adopted the *Julian day number* (often known as the *Julian day* or the *Julian date*) which started with zero at Greenwich mean noon on 1st January, 4713 BC, and most astronomical calculations begin by converting the given calendar date into the equivalent Julian day number. Note that the Julian day numbers begin at noon, and so are out of step by half a day with the civil calendar. For example, exactly 2 446 113 days had elapsed at noon on 16th February 1985 since noon on 1st January 4713 BC, so the Julian day number was 2 446 113 at that moment. By 6 am the following day, a further 0.75 days had passed, so the Julian day number was then 2 446 113.75.

AstroScript carries routines for converting between calendar dates given in the Gregorian calendar and Julian day numbers. These are based on the algorithms presented in Jean Meeus' excellent book *Astronomical Algorithms* (see *Bibliography*). The commands are *find_date(⟨string⟩)* and *display_date(⟨string⟩)*, where ⟨*string*⟩ is either 'julian' to convert from calendar to Julian dates and display the result, or 'calendar' for the reverse process. Thus the following AstroScript fragment would find and display the Julian day number corresponding to a given Gregorian date:

```
find_date(julian)
display_date(julian)
```

The disk which comes with this book has a more flexible AstroScript program called *calendar.txt* which allows you to convert in either direction. It is as follows:

```
{Clear the screen and display a title at the top}
display_title(++ DATE CONVERSIONS ++)

{Convert to Julian?}
ask_for_character(Convert from calendar date to Julian,yn,c1)
if (c1=Y)
   find_date(julian)
   display_date(julian)
end_if

{Convert to calendar?}
if (c1!Y)
   ask_for_character(....... from Julian date to calendar,yn,c2)
   if (c2=Y)
      find_date(calendar)
      display_date(calendar)
      display_time(solar)
   end_if
end_if

{Another calculation?}
new_line
ask_for_repeat
```

The conversion algorithms are valid for any zero or positive Julian day, i.e. for any date on or after 1.5 January 4713 BC, but will not work for dates before then. If you try to convert an illegal date, you will get the error message ++ *illegal date*, followed by indications from the internal procedures which detected the problem. The same message will also be seen if you try to convert any date which has a year of zero, or which lies between the 5th and 14th October 1582 inclusive. Here is an example of running *calendar.txt*:

```
                         ++ DATE CONVERSIONS ++
                         ----------------------

       Convert from calendar date to Julian ......... ? Y
       Calendar date (d,m,y; BC neg) ................ ? 19,5,1950
       Local civil time (h,m,s) ..................... ? 23,30,0
       Time zone (h ahead of UT; W negative) ........ ? 0
       Daylight saving (h ahead of zone t) .......... ? 1
       The Julian date is ...........................    2433421.4375

       Again (Y or N) ............................... ? Y

       Same date and time (Y or N) .................. ? Y
       Same time zone (Y or N) ...................... ? Y

                         ++ DATE CONVERSIONS ++
                         ----------------------

       Convert from calendar date to Julian ......... ? n
       ....... from Julian date to calendar ......... ? Y
       The local calendar date is ...................    Friday 19-May-1950
       The universal time is (h,m,s) ................    22 30  0.0000
       The local civil time is (h,m,s) ..............    23 30  0.0000
       Estimated ephemeris/dynamic time is (h,m,s) ..    22 30 27.9193

       Again (Y or N) ............................... ? n
```

Notice that an estimate of the dynamical time is displayed wherever possible by the *display_time* command.

# The celestial sphere

When you look at the sky from a point on the Earth, it is as if you are looking at the inside surface of a huge sphere surrounding you, rather like being inside a planetarium, on which are fixed the Sun, Moon, planets, stars, and other celestial objects. This sphere seems to be rotating steadily, making one complete revolution in about 24 hours. Of course, we know that the heavenly bodies are not all fixed to a sphere at the same distance from us but are really scattered across the vastness of the Universe. Nevertheless, the illusion is a useful one since it helps us to make astronomical calculations.

Imagine that there is an observer on the surface of the Earth in the northern hemisphere looking towards the south. He sees the sky around him as the inside surface of the celestial sphere. Now imagine that you are outside that sphere looking in towards the Earth. Figure 3 illustrates what you would see. The Earth (with the observer on it) is the small dot marked O in the centre of the sphere. The plane of the observer's *horizon*, extended until it cuts the celestial sphere, is marked Hor, with its cardinal points north, east, south, west marked N, E, S, and W, respectively. The point directly overhead, Z, is the observer's *zenith*, and the point, D, directly underneath on the opposite side of the sphere is the *nadir*.

You can now imagine that several other curves have been inscribed on the surface of the sphere. We have already mentioned the horizon which is an example of a *great circle*, that is a circle on the sphere whose centre is also at the centre of the sphere. Such circles are the largest that can be drawn on the sphere's surface. Any other circle, whose centre does not coincide with that of the sphere, must have a smaller diameter and is called a *small circle*. The horizon, then, is a great circle. Another great circle is the curve drawn through the points Z, B, S, D, N, and P. This is the observer's *meridian*. It passes through the zenith point, and crosses the horizon at right angles at its southern point. (Note that we are discussing the situation for an observer in the northern hemisphere. We shall look at the differences for southern observers in a moment.) A third great circle is also drawn on the diagram. Imagine that the plane of the Earth's equator is extended until it intersects the surface of the celestial sphere. The curve of intersection is this third great circle, marked Eq. It passes through the E and W points, and is called the *celestial equator*. The angle, $\phi$, between the directions of the zenith and the southern point of the celestial equator (i.e. the angle BOZ) is dependent on the observer's position on the Earth, usually described in terms of the *geographical longitude* and *geographical latitude*. The angle $\phi$ is equal to the geographical latitude. If you extend one of your arms towards the southernmost point of the equator, and your

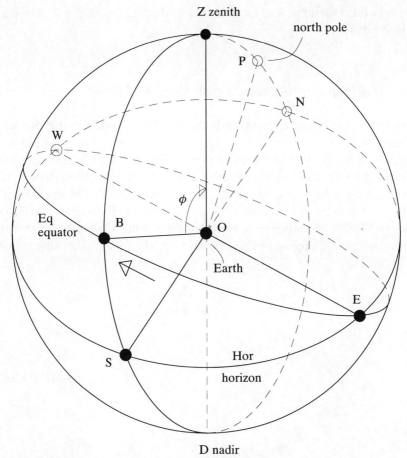

Figure 3. The celestial sphere for a northern observer.

other arm towards your zenith, the angle between them will be your geographical latitude.

Every great circle has a corresponding pair of points called its *poles* which are equidistant from all points on the great circle itself, and are as far apart from each other as possible. Thus the zenith, Z, and nadir, D, are the poles of the horizon, whilst the east and west points, E and W, are the poles of the observer's meridian. The northern pole of the celestial equator is marked P in the diagram, and is the point of intersection with the celestial sphere of the rotation axis of the Earth extended out from the north pole. As the Earth spins about this axis, so it gives the illusion that the celestial sphere itself is spinning about the (stationary) Earth in the opposite direction. 'Fixed' stars seem to us to be describing small circles centred on the point P, moving in such a direction that they cross the meridian from east to west at their southernmost points (direction of the arrow). Stars near enough to the pole stay above the horizon all the time, whilst others spend part of the time below the horizon, *rising* in the east and later *setting* in the west.

## Horizon coordinates: azimuth and altitude

The position of a heavenly body is always described in terms of a pair of *coordinates*, defined with respect to a reference great circle. From the point of view of the observer on the Earth's surface, it is convenient to use the horizon as the reference circle (see Figure 4). Imagine that a great circle, ZXX′, has been drawn through the zenith and the object at X, which cuts the horizon at point X′. The *azimuth* of X is then the angle $A$ subtended at O between the north point of the horizon and X′, measured in the sense NESW. The *altitude* or *elevation* of X is the angle $a$ between the points X and X′ subtended at O, measured positive towards the zenith. Thus the azimuths of the

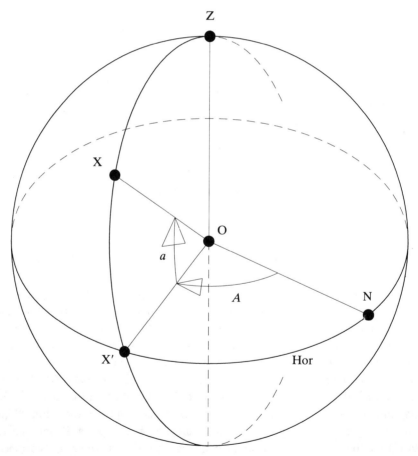

Figure 4. Horizon coordinates: azimuth, $A$, and altitude or elevation, $a$.

north, east, south, and west points of the horizon are 0, 90, 180, and 270 degrees, respectively, whilst the altitudes of the zenith and nadir are 90 and −90 degrees, respectively.

The observer's view of the sky is shown schematically in Figure 5, with the four parts representing views towards the south, north, east, and west. Consider first the view to the south. The horizontal line along the bottom represents the observer's horizon with the south point in the centre and the east and west points to the left and right, respectively. The vertical line from S to Z is the observer's meridian, and is a line of constant azimuth (180 degrees in this case). The curves extending down to the horizon from Z (such as ZE) are also curves of constant azimuth with values indicated by the numbers along the bottom below the horizon. The horizontal lines marked 30 and 60 are lines of constant altitude (30 and 60 degrees, respectively). The curve labelled Eq marks the trace of the celestial equator, and the curve above it, St, shows the path of a star, moving in the direction of the arrow and returning to approximately the same position in the sky after 24 hours.

The other views show the same features as seen when looking towards the other cardinal points of the horizon. Note that this representation makes the sky look much smaller than it really is. If you were the observer facing due south, the east point would be over your left shoulder, and the west point over your right shoulder. The zenith would be the point directly over your head. Each view in the figure therefore

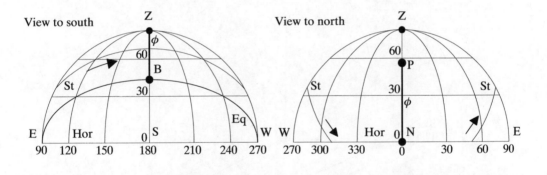

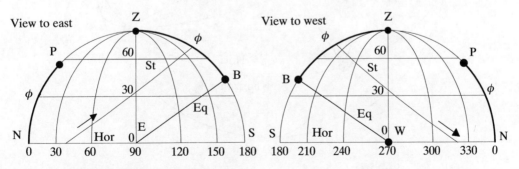

Figure 5. Views of the northern sky.

represents half the visible sky which has been compressed onto the flat page by means of a particular *mapping transformation* (see the chapter called *Sky displays*).

## Southern observers

Observers in the Earth's southern hemisphere see much the same features, but with the important difference that the sky appears to revolve in the opposite direction. Figure 6 is similar to Figure 3 but illustrates the southern celestial sphere as if you were outside it looking in (as on Figure 3). The observer in the southern hemisphere is on the Earth at the centre, O, of the sphere looking towards the north. The angle between the zenith and the northern point of the equator is the observer's geographical latitude *south*, usually designated by a negative number. Thus, if $\phi'$ is the magnitude of the southern latitude then $\phi'$ is the angle between Z and the northern point of the equator. The point P' is the south celestial pole, around which the stars

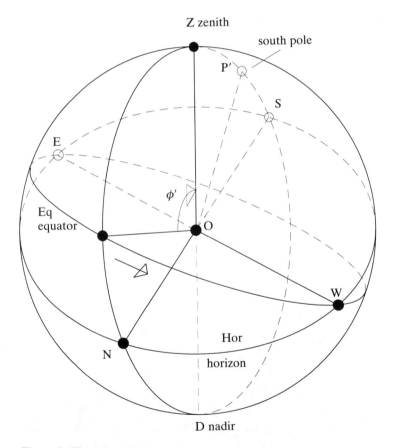

Figure 6. The celestial sphere for a southern observer.

appear to revolve. Views of the sky seen by the observer on the ground looking towards the cardinal points are represented in Figure 7.

## Displaying the sky

You can use the AstroScript command *compute_sky* to obtain views of the sky similar to those shown in Figures 5 and 7 for observers looking in any direction in any part of the world. The command is described in detail in the chapter *Sky displays*, but you can run it straight away using the script file *ast_desk.txt* supplied on the disk which is as follows:

> {Clear the screen, and display a title}
> display_title(+++ Astro Desk +++)
>
> {Invoke the astro-desk command}
> compute_sky
>
> {Another go?}
> ask_for_repeat

Just keep on pressing the *enter* or *return* key in response to every question. The default values (as defined in the file *setup.dat*) will then be used. Eventually, you will see a

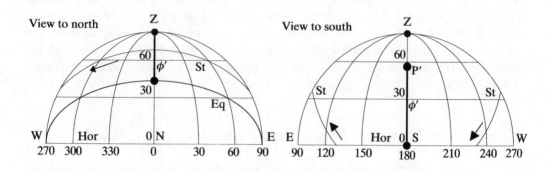

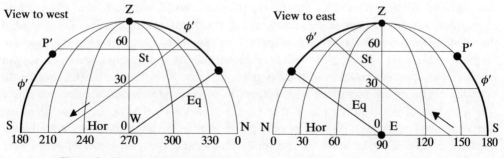

Figure 7.  Views of the southern sky.

multicoloured tabular display, updated every second (if your computer can go that fast) showing the positions of the Sun, Moon, and major planets, something like this:

```
System      G Sidereal Time  L Sidereal Time  Universal Time   Local Civil Time
time            3 58 34.9        3 58 43.3       15 44  7.0        15 44  7.0


Geographic longitude (d,m,s)  0 2 5.7         Julian Date      Local Civil Date
Geographic latitude (d,m,s) ? 52 10 12.0                           Sunday
Time zone (h ahead of UT; W negative) 0     J  2449803.81        26-Mar-1995
Daylight saving (h ahead of zone t) ? 0


                                              SOLAR SYSTEM OBJECTS: page    1


OBJECT   RA (H,M,S)   DEC (D,M,S)    AZ (D,M,S)      ALT (D,M,S) DIST (AU) PHASE
Sun       0 20 14.9      2 11 22.3   241 58 46.6     22 38 46.9  0.997615
Moon     20 48 32.1  - 12 58 56.1   275 47 33.2  - 20 57 13.7 378753 km  0.21
Mercury  23 20 33.7  -  6 45 50.3   249 34 31.6      6 53 12.8  1.263185  0.86
Venus    22  1 58.1  - 12 37 35.1   261 32 40.9  -  9 26 48.0  1.194841  0.78
Mars      9  6 39.0    20  3 10.6    87 17 45.6     23 36 47.7  0.856998  0.94
Jupiter  16 56 41.0  - 21 49  8.1   334 15 38.1  - 57 39 10.5  4.932012  0.99
Saturn   23 16 52.0  -  6 34 43.2   250 26  9.5      6 30 19.7 10.601584  1.00
Uranus   20  8 27.8  - 20 41 33.6   278 34 52.4  - 32 59 40.1 20.095727  1.00
Neptune  19 48 27.6  - 20 32 34.5   283  2 37.1  - 35 53 31.3 30.496976  1.00
```

Now press the *D* key, read the table of instructions, and then press another key. After a pause (whose length depends upon the speed of your computer and whether or not you have a coprocessor), the screen will display a graphical representation of the sky with the positions of various celestial objects shown on it. By default, you will be looking directly along the ground due south at a point near the Greenwich meridian about 52° N latitude. The brown curve represents the trace of the *celestial equator*, the red curve the trace of the *plane of the ecliptic*, and the magenta curve the trace of the *galactic plane* (Milky Way). More details of the compute_sky command may be found in the *Sky graphics* section on page 96.

## *Equatorial coordinates: right ascension and declination*

The horizon coordinates, azimuth and altitude, are ideal for specifying the position of a celestial object as seen by an observer on the ground at a particular moment, but are not so good for comparing the positions of many different objects with each other, since they change constantly as the Earth rotates on its axis. The so-called 'fixed' stars appear to be stationary in space with respect to each other, largely because they are so far away, and it is convenient therefore to use a different coordinate system in which the apparent rotation of the sky plays no part. As with all celestial coordinate systems, positions are measured with respect to a fixed direction in a reference plane (the direction of north in the plane of the horizon in the case of horizon coordinates). The *equatorial coordinate system* uses the plane of the Earth's equator, extended until it cuts the celestial sphere, as its reference plane, and the *first point of Aries* (see below) as its fixed direction.

The celestial sphere showing equatorial coordinates is drawn in Figure 8. Imagine that you are an observer at O on the surface of the Earth. Your horizon is the great circle labelled Hor with its west and south points marked as W and S respectively. The great circle PZAS is your meridian, and the great circle labelled Eq through A, X′, and W is the celestial equator (marked Eq). The north pole of the equator is P, and Z is your zenith. Now imagine that you wish to know the equatorial coordinates of the star at X. Its meridian is the great circle through P and X, which crosses the equator at right angles at point˘X′. The *declination* of X is then the angle, $\delta$, the angle subtended at O by the points X and X′, measured positive towards the north pole and negative towards the south pole. (Note that angles can be represented, as here, by the lengths of the curves on the surface of the celestial sphere.) The *right ascension* of X is the angle $\alpha$, subtended at O by the points X′ and ♈ measured in an eastwardly direction. The point ♈ is the point on the celestial sphere where the planes of the equator and the ecliptic

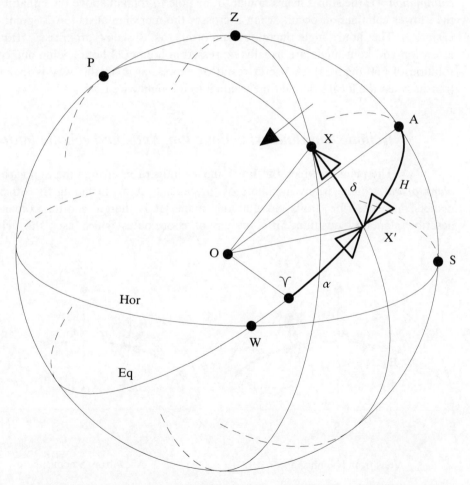

Figure 8. Equatorial coordinates: right ascension, $\alpha$, declination, $\delta$, and hour angle, $H$.

cross one another and it is called the *vernal equinox*, or *first point of Aries*. Note that the sense of measuring $\alpha$ is opposite to the apparent direction of motion of the object which is moving westwards in the diagram in the direction shown by the arrow near $X$.

Figure 9 shows the observer's view of the same thing as seen from the ground both in the northern hemisphere and in the southern hemisphere. As the day progresses, so both the points $X'$ and $\Upsilon$ move along the equator at the same rate. Furthermore, the track of the star is a small circle parallel to the equator. Thus both $\alpha$ and $\delta$ remain constant, and are independent of the Earth's rotation and the observer's viewpoint. Also marked on both Figures 8 and 9 is the *hour angle* of $X$. This is the angle, $H$, between A and $X'$, measured along the equator in the same sense as the apparent motion of X (and hence in the opposite sense to that of $\alpha$). A is the point where the celestial equator crosses the observer's meridian. The hour angle is zero as the star crosses the observer's meridian moving westerly; this event is known as *transit* or *upper culmination*. (If the star is near enough to the pole to remain above the equator all day, then lower culmination occurs when it crosses the meridian at its lowest point moving easterly.) The hour angle increases uniformly as the day progresses. Both right ascension and hour angle are usually expressed in hours (24 hours being one complete rotation of 360 degrees). A useful result of expressing it in this way is that the star transits when the local sidereal time is equal to the right ascension.

## Ecliptic coordinates: ecliptic longitude and ecliptic latitude

The plane in which the Earth moves in its orbit around the Sun is called *the plane of the ecliptic*. Many members of the Solar System, including the other major planets, also describe orbits close to this plane. It is therefore often convenient to describe planetary positions in a system of coordinates which uses the ecliptic as

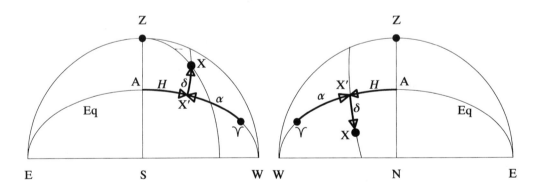

View from N hemisphere                    View from S hemisphere

Figure 9. Equatorial coordinates seen from the ground, both in the northern hemisphere and in the southern hemisphere.

reference. Figure 10 shows the celestial sphere with the planes of the horizon, equator, and ecliptic drawn as great circles Hor, Eq, and Ecl respectively. The ecliptic is inclined at an angle of about 23° 26′ to the equator which is also the angle made by the Earth's north–south axis with respect to the perpendicular to the ecliptic. This angle, $\varepsilon$, is known as the *obliquity of the ecliptic*. The point K is the north pole of the ecliptic, and V is the planet whose ecliptic coordinates we wish to define. Let KVV′ be the great circle through K and V; it intersects the ecliptic at V′. Then the ecliptic latitude, $\beta$, is the angle V′OV, measured positive towards the north pole of the ecliptic (the same side as the north pole of the equator) and negative towards the south pole. The ecliptic longitude, $\lambda$, is the angle, ϒOV′ measured in the same sense as the right ascension i.e. in the opposite direction to the diurnal motion of V.

The situation as viewed from the ground in the northern and southern hemispheres is shown in Figure 11. Both $\lambda$ and $\beta$ are usually measured in degrees. During the course of the year the Sun appears to progress along the trace of the ecliptic in the sense NWSE. Its ecliptic latitude is always zero (to a very good approximation, perturbed

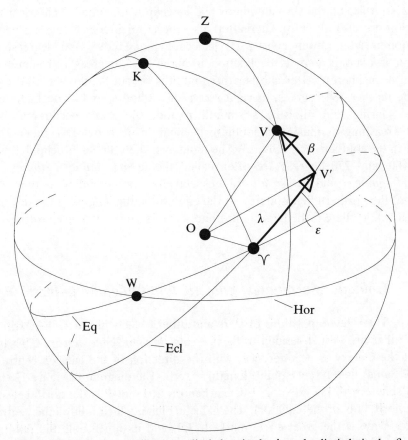

Figure 10. Ecliptic coordinates: ecliptic longitude, $\lambda$, and ecliptic latitude, $\beta$.

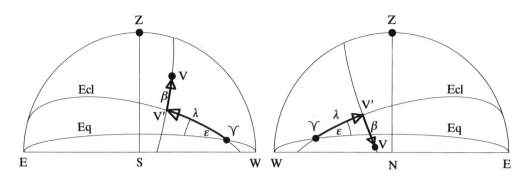

View from N hemisphere                    View from S hemisphere

Figure 11. Ecliptic coordinates seen from the ground, both in the northern hemisphere and in the southern hemisphere.

only slightly by the influences of the other bodies in the Solar System). On or about March 21st, it lies on the vernal equinox and its ecliptic longitude, right ascension, and declination are also all zero. Thereafter, its ecliptic longitude increases steadily until three months later (about June 21st) its ecliptic longitude is 90 degrees, its right ascension is 6 hours and its declination is about 23.5 degrees (midsummer in the northern hemisphere). After another three months (about September 21st) its ecliptic longitude reaches 180 degrees, its right ascension 12 hours, and its declination returns to zero. During the following six-month period, the Sun's declination is always negative, reaching its maximum southern declination on about December 21st (mid-summer in the southern hemisphere). The moments when the Sun's ecliptic longitude is 0, 90, 180, and 270 degrees are often called the *vernal equinox*, *summer solstice*, *autumnal equinox*, and *winter solstice*, respectively, corresponding as they do to the seasons of the northern hemisphere. At the equinoxes, the lengths of the day and night are equal, whilst there is the largest difference between the two at each solstice.

## Galactic coordinates: galactic longitude and galactic latitude

The relative positions of stars and other objects in our Galaxy with respect to each other are best described using the *galactic coordinate system*. This takes the plane of the Galaxy as its reference, with the direction of the galactic centre as seen from the Sun as its reference point, longitude zero. The diagram in Figure 12 shows the celestial sphere with the equator, Eq, marked on it. Point P is the north celestial pole (i.e. the north pole of the equator). The galactic plane, which follows the centre line of the Milky Way, is the great circle marked Gpl, and its north pole the point marked NGP. The point GC is the direction of the galactic centre as seen from the Sun. If X is

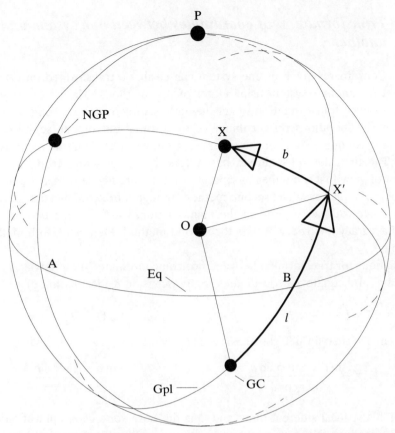

Figure 12.  Galactic coordinates: galactic longitude, $l$, and galactic latitude, $b$.

the star whose coordinates we wish to define then we draw a great circle through NGP and X which cuts the galactic plane at right angles at X'. The galactic latitude of X is the angle, $b$, subtended at O by X' and X, positive if north of the galactic plane and negative if south of it. The galactic longitude is the angle, $l$, measured along the galactic plane subtended at O by GC and X', in the anticlockwise sense when viewed from the north galactic pole looking down onto the plane.

The directions of NGP and GC have been defined by the International Astronomical Union in terms of 1950 coordinates (see the section *precession* for an explanation of how the right ascension and declination change slowly with time). In these terms the north galactic pole has right ascension $12^h$ $49^m$ ($192°.25$), and declination $+27°.4$. Another great circle through P and NGP cuts the equator perpendicularly at A and this is therefore $192°.25$ round the equator from the first point of Aries. The galactic plane cuts the equator at B, a further 90 degrees round from A. Point B is the *rising node* of the galactic plane on the equator. The galactic centre direction is defined to be $33°$ distant from it (i.e. the angle subtended at O by GC and B is $33°$).

## Transformation of coordinates between one system and another

Coordinates given in one system can easily be transformed into the corresponding set for another system using either of two methods. The first is to apply the equations for each transformation to get the answer directly. The second is to perform a set of matrix operations on a column vector formed out of one set to obtain the corresponding column vector for the other set, from which the coordinates may then be obtained. The first method is direct, but suffers the disadvantage from the programmer's point of view that a number of separate routines are needed, each manipulating a separate pair of equations. The second method, though more complicated, nevertheless can use a single routine to perform the manipulations, and this can be optimised for speed and accuracy. AstroScript uses the second method. Details of both methods are given below.

The equations for transforming between horizon coordinates (hour angle $H$, azimuth $A$, altitude $a$) and equatorial coordinates (right ascension $\alpha$, declination $\delta$) are

$$H = \text{LST} - \alpha, \qquad\qquad \alpha = \text{LST} - H,$$

$$\sin a = \sin \delta \sin \phi + \cos \delta \cos \phi \cos H, \quad \sin \delta = \sin a \sin \phi + \cos a \cos \phi \cos A,$$

$$\cos A = \frac{\sin \delta - \sin \phi \sin a}{\cos \phi \cos a}, \qquad \cos H = \frac{\sin a - \sin \phi \sin \delta}{\cos \phi \cos \delta}.$$

where LST is the local sidereal time, and $\phi$ is the observer's geographical latitude (N positive, S negative). Note that whenever equations like these are used to obtain a longitude-like coordinate (hour angle, azimuth, right ascension etc.), there is always an ambiguity on taking inverse sin, cos, or tan. For example, if $x$ is the result of an inverse tangent operation, then the answer can be either $x$ or $180° + x$. In the cases of the equations above, you can remove the ambiguity by taking $\sin H$ on the left, or $\sin A$ on the right. If that is negative, then $A$ on the left or $H$ on the right needs no correction. Otherwise, subtract $A$ on the left or $H$ on the right from 360°. On other occasions, it may be convenient to express the result as the inverse tangent of a fraction $y/x$. Your computer language may then have a function like ATAN2 which automatically resolves the ambiguity for you. Otherwise you may need to add or subtract 180° to your result to bring it within the correct quadrant as follows:

    0 to 90:     $x$ and $y$ both positive
    90 to 180:   $x$ negative, $y$ positive
    180 to 270:  $x$ and $y$ both negative
    270 to 360:  $x$ positive, $y$ negative.

You need be worried by none of this when using AstroScript, as all such details are dealt with in the background and kept strictly out of your sight.

Transformations between equatorial and ecliptic coordinates are governed by the following equations in which $\lambda$ represents the ecliptic longitude, $\beta$ the ecliptic latitude, and $\varepsilon$ the obliquity of the ecliptic:

$$\sin \delta = \sin \beta \cos \varepsilon + \cos \beta \sin \varepsilon \sin \lambda, \qquad \sin \beta = \sin \delta \cos \varepsilon - \cos \delta \sin \varepsilon \sin \alpha,$$

$$\tan \alpha = \frac{\sin \lambda \cos \varepsilon - \tan \beta \sin \varepsilon}{\cos \lambda}, \qquad \tan \lambda = \frac{\sin \alpha \cos \varepsilon + \tan \delta \sin \varepsilon}{\cos \alpha}.$$

The obliquity of the ecliptic changes slowly with time. Its mean value is given approximately by the expression

$$\varepsilon = 23° \, 26' \, 21''.448 - 46''.8150T - 0''.00059T^2 + 0''.001813T^3$$

where $T$ is the number of Julian centuries of $36\,525$ days from the epoch J2000.0 (midday on the Greenwich meridian on January 1st in the year 2000 – strictly by barycentric dynamical time, TDB, but UT will usually do). The error in this expression is always less than $1''$ for the period of 2000 years either side of J2000.0. The true value of the obliquity is the mean value plus a small correction for nutation in obliquity caused by the wobbling motion of the Earth. This is handled automatically within AstroScript.

The equations for transforming between equatorial and galactic coordinates are:

$$\sin \delta = \cos b \cos 27.4 \sin (l - 33) + \sin b \sin 27.4,$$

$$\tan (\alpha - 192.25) = \frac{\cos b \cos (l - 33)}{\sin b \cos 27.4 - \cos b \sin 27.4 \sin (l - 33)},$$

$$\sin b = \cos \delta \cos 27.4 \cos (\alpha - 192.25) + \sin \delta \sin 27.4,$$

$$\tan (l - 33) = \frac{\sin \delta - \sin b \sin 27.4}{\cos \delta \sin (\alpha - 192.25) \cos 27.4}$$

where $l$ represents the galactic longitude and $b$ the galactic latitude. Strictly, the right ascension and declination should be corrected for precession to the standard equinox B1950.0 before using these equations. (B1950.0 is approximately the instant 0.9,1,1950.)

Matrix coordinate conversions require first that the given pair of coordinates is converted into a column vector of three values. Next this column vector is multiplied by one or more matrices to convert from one system to another, perhaps going via an intermediate system. Finally, the resulting column vector is converted back into a pair of coordinates. If the given coordinates are represented by the symbols $(\mu, \nu)$, where $\mu$ is the longitude-like coordinate (hour angle, azimuth, right ascension, ecliptic longitude, galactic longitude), and $\nu$ is the latitude-like coordinate (altitude, declination, ecliptic latitude, galactic latitude), then the three-element column vector, $\mathbf{v}$, is defined to be

$$\mathbf{v} = \begin{pmatrix} x \\ y \\ z \end{pmatrix} = \begin{pmatrix} \cos \mu \cos \nu \\ \sin \mu \cos \nu \\ \sin \nu \end{pmatrix}.$$

The converted column vector, **w**, is obtained by pre-multiplying **v** by one or more $3 \times 3$ matrices according to which transformations you wish to make. Thus if **A** is one such matrix, then

$$\mathbf{w} = \begin{pmatrix} m \\ n \\ p \end{pmatrix} = \mathbf{A} \cdot \mathbf{v} = \begin{pmatrix} a & b & c \\ d & e & f \\ g & h & i \end{pmatrix} \cdot \begin{pmatrix} x \\ y \\ z \end{pmatrix}.$$

The resulting pair of coordinates, $(\theta, \psi)$, can then be obtained from **w** by

$$\tan \theta = n/m,$$

$$\sin \psi = p,$$

where $\theta$ is the longitude-like coordinate and $\psi$ is the latitude-like coordinate.

The matrices **A** are as follows:

$$\begin{pmatrix} m \\ n \\ p \end{pmatrix}_{H,\delta} = \begin{pmatrix} -\sin\phi & 0 & \cos\phi \\ 0 & -1 & 0 \\ \cos\phi & 0 & \sin\phi \end{pmatrix} \cdot \begin{pmatrix} x \\ y \\ z \end{pmatrix}_{A,a}$$

$$\begin{pmatrix} m \\ n \\ p \end{pmatrix}_{A,a} = \begin{pmatrix} -\sin\phi & 0 & \cos\phi \\ 0 & -1 & 0 \\ \cos\phi & 0 & \sin\phi \end{pmatrix} \cdot \begin{pmatrix} x \\ y \\ z \end{pmatrix}_{H,\delta}$$

$$\begin{pmatrix} m \\ n \\ p \end{pmatrix}_{H,\delta} = \begin{pmatrix} \cos\mathrm{LST} & \sin\mathrm{LST} & 0 \\ \sin\mathrm{LST} & -\cos\mathrm{LST} & 0 \\ 0 & 0 & 1 \end{pmatrix} \cdot \begin{pmatrix} x \\ y \\ z \end{pmatrix}_{\alpha,\delta}$$

$$\begin{pmatrix} m \\ n \\ p \end{pmatrix}_{\alpha,\delta} = \begin{pmatrix} \cos\mathrm{LST} & \sin\mathrm{LST} & 0 \\ \sin\mathrm{LST} & -\cos\mathrm{LST} & 0 \\ 0 & 0 & 1 \end{pmatrix} \cdot \begin{pmatrix} x \\ y \\ z \end{pmatrix}_{H,\delta}$$

$$\begin{pmatrix} m \\ n \\ p \end{pmatrix}_{\lambda,\beta} = \begin{pmatrix} 1 & 0 & 0 \\ 0 & \cos\varepsilon & \sin\varepsilon \\ 0 & -\sin\varepsilon & \cos\varepsilon \end{pmatrix} \cdot \begin{pmatrix} x \\ y \\ z \end{pmatrix}_{\alpha,\delta}$$

$$\begin{pmatrix} m \\ n \\ p \end{pmatrix}_{\alpha,\delta} = \begin{pmatrix} 1 & 0 & 0 \\ 0 & \cos\varepsilon & -\sin\varepsilon \\ 0 & \sin\varepsilon & \cos\varepsilon \end{pmatrix} \cdot \begin{pmatrix} x \\ y \\ z \end{pmatrix}_{\lambda,\beta}$$

$$\begin{pmatrix} m \\ n \\ p \end{pmatrix}_{l,b} = \begin{pmatrix} -0.066\,988\,739\,415\,151 & -0.872\,755\,765\,851\,993 & -0.483\,538\,914\,632\,184 \\ 0.492\,728\,466\,075\,323 & -0.450\,346\,958\,019\,961 & 0.744\,584\,633\,283\,031 \\ -0.867\,600\,811\,151\,435 & -0.188\,374\,601\,722\,920 & 0.460\,199\,784\,783\,852 \end{pmatrix} \cdot \begin{pmatrix} x \\ y \\ z \end{pmatrix}_{\alpha,\delta}$$

$$\begin{pmatrix} m \\ n \\ p \end{pmatrix}_{\alpha,\delta} = \begin{pmatrix} -0.066\,988\,739\,415\,151 & 0.492\,728\,466\,075\,323 & -0.867\,600\,811\,151\,435 \\ -0.872\,755\,765\,851\,993 & -0.450\,346\,958\,019\,961 & -0.188\,374\,601\,722\,920 \\ -0.483\,538\,914\,632\,184 & 0.744\,584\,633\,283\,031 & 0.460\,199\,784\,783\,852 \end{pmatrix} \cdot \begin{pmatrix} x \\ y \\ z \end{pmatrix}_{l,b}$$

The matrices for conversion between equatorial coordinates and galactic coordinates are best expressed in terms of numbers as above, for they do not depend on place or time. You might think it unnecessary to use so many places of decimals. This is true for most conversions, but the accuracy is greatly improved near the poles by increasing the precision of the matrix elements.

## *Converting coordinates using AstroScript*

AstroScript provides you with a powerful command for converting any pair of coordinates given in one system into the corresponding pair given in any other system. With this you can convert the azimuth and altitude of a star into its galactic longitude and latitude, or its right ascension and declination into its ecliptic longitude and latitude etc. The command is called *convert_coordinates*, and it takes two single-character arguments to direct its action. The first argument specifies which system the given coordinates belong to, and the second argument specifies into which system they are to be transformed. Each may be set to one of 'h', 'r', 'a', 'e', or 'g' (but see also the special cases of 's', 'm', 'p', and 'o' in the section *Calculation commands* of *Appendix A*) corresponding to hour angle/declination, right ascension/declination, azimuth/altitude, ecliptic longitude/latitude, and galactic longitude/latitude, respectively. Hence to convert from azimuth/altitude into, say, galactic longitude/latitude you would write *convert_coordinates(a,g)*. Of course, you need to be able to type in your coordinates from the keyboard, and then to display the result of the conversion. Two other AstroScript commands allow you to do this. They are *ask_for_coordinates* and *display_coordinates*. Both take a single-character argument to specify the coordinates to be accepted or displayed. Thus a basic coordinate conversion program would be

```
ask_for_coordinates(a)
convert_coordinates(a,g)
display_coordinates(g)
```

which will carry out the conversion from the horizon to the galactic system.

Provided on the disk which accompanies this book is a more generalised AstroScript transformation program called *gencon.txt*. Type in *ascript gencon.txt* to run it. This is what it contains:

```
{Clear the screen and display a title at the top}
display_title(++ TRANSFORMATION of CELESTIAL COORDINATES ++)

{Set the delimiter to % so that commas can be used in the prompt string}
set_delimiter(%)

{Get the coordinates to transform from and into}
ask_for_character(Transform from which system [h,a,r,e,g]%hareg%c1)
ask_for_character(.......... to which system [h,a,r,e,g]%hareg%c2)
new_line
```

```
{Write an appropriate message depending on the coordinates}
if (c1=h)
   write_string(Please input hour angle and declination:)
end_if
if (c1=g)
   write_string(Please input galactic longitude and latitude:)
end_if
if (c1=r)
   write_string(Please input right ascension and declination:)
end_if
if (c1=a)
   write_string(Please input azimuth and altitude:)
end_if
if (c1=e)
   write_string(Please input ecliptic longitude and latitude:)
end_if
new_line

{Get the coordinates themselves which are to be transformed...}
ask_for_coordinates(c1)

{... and transform them}
convert_coordinates(c1%c2)

{Set the text colour to red...}
set_normal_text(4)

{... and display the result}
new_line
display_coordinates(c2)

{Reset the text colour to its normal state}
set_normal_text(7)
new_line

{Ask if we want to do it again}
ask_for_repeat
```

Note that we wish to use commas in the prompt strings in this program, so we have to change the delimiter character from comma to the percent symbol, %. Here is an example of running this program:

```
           ++ TRANSFORMATION of CELESTIAL COORDINATES++
           --------------------------------------------

Transform from which system [h,a,r,e,g] ...... ? r
.......... to which system [h,a,r,e,g] ...... ? a

Please input right ascension and declination:
Right-ascension (h,m,s) .................... ? 23,34,45.67
Declination     (d,m,s) .................... ? -56,10,23
```

```
Calendar date (d,m,y; BC neg) ................ ? 26/ 3/ 1995
Local civil time (h,m,s) ..................... ? 17 14 26.0000
Time zone (h ahead of UT; W negative) ........ ? 7
Daylight saving (h ahead of zone t) .......... ? 0
Geographic longitude (d,m,s) ................. ? 120,0,0
Geographic latitude (d,m,s) .................. ? 45,34,0

The azimuth is (d,m,s; N=0) ..................    227 52  0.6659
The altitude is (d,m,s) ......................  -  43  4 15.5644

Again (Y or N) ............................... ? Y

Same date and time (Y or N) .................. ? Y
Same time zone (Y or N) ...................... ? Y
Same geographical longitude (Y or N) ......... ? Y
Same geographical latitude (Y or N) .......... ? Y

              ++ TRANSFORMATION of CELESTIAL COORDINATES ++
              ----------------------------------------------

Transform from which system [h,a,r,e,g] ...... ? a
.......... to which system [h,a,r,e,g] ...... ? r

Please input azimuth and altitude:
Azimuth  (d,m,s; N=0) ........................ ? 227 52 0.6659
Altitude (d,m,s) ............................. ? -43 4 15.5644

The right ascension is (h,m,s) ...............    23 34 45.6700
The declination is (d,m,s) ...................  -  56 10 23.0000

Again (Y or N) ............................... ? n
```

If you wish to use *convert_coordinates* to transform to or from the galactic coordinate system, you should recognise that the transformation matrix is correct for the epoch B1950.0, so you should first carry out a correction for precession (see next section) on the right ascension and declination. Your AstroScript program would therefore look something like this:

```
ask_for_coordinates(r)
correct_for_precession
convert_coordinates(r,g)
display_coordinates(g)
```

or, for transformations in the other direction:

```
ask_for_coordinates(g)
convert_coordinates(g,r)
correct_for_precession
display_coordinates(r)
```

but in many cases the correction for precession makes only a small difference and can be ignored.

## *Precession*

Anyone who has played with a spinning top as a child will know that its axis slowly gyrates about the vertical direction as it spins. This phenomenon is common to any spinning object which is under the influence of an external couple, and is called *gyroscopic precession*. In the case of the spinning top, the couple is provided by the weight of the top which pulls vertically downwards through its centre of gravity. The Earth is also a spinning object and it, too, exhibits the same precessional behaviour. The gravitational attractions of the Moon and the Sun, acting on the equatorial bulges of the slightly non-spherical Earth, cause the couple which results in the north–south axis precessing slowly about a line through the centre of the Earth and perpendicular to the plane of the ecliptic with a period of about 26 000 years. The effect of this *luni-solar precession* is to make the direction of the line of intersection of the rising node of the celestial equator on the plane of the ecliptic (the first point of Aries or vernal equinox) move steadily at a rate of about 50 arcseconds per year.

There is also another, smaller, precession from the gravitational influences of the planets. If the celestial equator were fixed (i.e. not perturbed by luni-solar precession), their combined effect would be to cause the equinox to move by about 12 arcseconds per century, and to decrease the obliquity of the ecliptic by about 47 arcseconds per century. This is called *planetary precession*. Both luni-solar and planetary precession are usually calculated together as *general precession*.

The positions of 'fixed' stars and other celestial objects are often defined by their right ascensions and declinations, or by their ecliptic longitudes and latitudes. In either case, the reference direction is that of the first point of Aries, so general precession causes these coordinates to change slowly with time. Hence it is necessary to specify the moment, or *epoch*, at which a particular pair of coordinates is valid. A correction can then be applied to convert the coordinates to values which are valid at another epoch. AstroScript incorporates an algorithm for performing this correction for general precession rigorously. The AstroScript command is *correct_for_precession* and it converts the currently held right ascension and declination into their new values. It asks you to supply the two precessional epochs if you have not already done so.

The AstroScript algorithm for general precession follows that laid out in the *Explanatory Supplement to the Astronomical Almanac* (see *Bibliography*). If $(\alpha_1, \delta_1)$ are the right ascension and declination valid at epoch $\varepsilon_1$, the column vector $\mathbf{r}_1$ is first obtained by

$$\mathbf{r}_1 = \begin{pmatrix} \cos \alpha_1 \cos \delta_1 \\ \sin \alpha_1 \cos \delta_1 \\ \sin \delta_1 \end{pmatrix}.$$

This is then multiplied by the precession matrix, $\mathbf{P}$, to form the new column vector, $\mathbf{r}_2$, appropriate for the new epoch $\varepsilon_2$ as follows:

$$\mathbf{r}_2 = \mathbf{P} \cdot \mathbf{r}_1.$$

If the components of $\mathbf{r}_2$ are

$$\mathbf{r}_2 = \begin{pmatrix} m \\ n \\ p \end{pmatrix},$$

the new coordinates, $(\alpha_2, \delta_2)$, are obtained from $\mathbf{r}_2$ by the equations

$$\tan \alpha_2 = n/m, \text{ and } \sin \delta_2 = p.$$

The precession matrix $\mathbf{P}$ is given by

$$\mathbf{P} = \begin{pmatrix} cx \cdot ct \cdot cz - sx \cdot sz & -sx \cdot ct \cdot cz - cx \cdot sz & -st \cdot cz \\ cx \cdot ct \cdot sz + sx \cdot cz & -sx \cdot ct \cdot sz + cx \cdot cz & -st \cdot sz \\ cx \cdot st & -sx \cdot st & ct \end{pmatrix},$$

where $cx = \cos \zeta_A$, $sx = \sin \zeta_A$, $cz = \cos z_A$, $sz = \sin z_A$, $ct = \cos \theta_A$, and $st = \sin \theta_A$. These arguments are calculated from the following expressions involving the interval in Julian centuries of 36 525 days between $\varepsilon_0$, the fundamental epoch J2000.0 (1.5 Jan. 2000), and the two epochs $\varepsilon_1$ and $\varepsilon_2$, where

$$T = [\text{JD}(\varepsilon_1) - \text{JD}(\varepsilon_0)]/36\,525,$$

$$t = [\text{JD}(\varepsilon_2) - \text{JD}(\varepsilon_1)]/36\,525,$$

$\text{JD}(\varepsilon)$ represents the Julian day number of epoch $\varepsilon$, and $\text{JD}(\varepsilon_0) = 2\,451\,545.0$,

$$\zeta_A = (2306.2181 + 1.396\,56T - 0.000\,139T^2)t + (0.301\,88 - 0.000\,344T)t^2 + 0.017\,998t^3$$

$$z_A = (2306.2181 + 1.396\,56T - 0.000\,139T^2)t + (1.094\,68 + 0.000\,066T)t^2 + 0.018\,203t^3$$

$$\theta_A = (2004.3109 - 0.853\,30T - 0.000\,217T^2)t + (-0.426\,65 - 0.000\,217T)t^2 - 0.041\,833t^3.$$

$\zeta_A$, $z_A$, and $\theta_A$ are all given in arcseconds by the above expressions.

There is an AstroScript program on the disk which comes with the book called *precess.txt*. It allows you to convert a pair of coordinates given in any system at one epoch into the corresponding pair at another epoch, going via the right ascension and declination to which the correction for precession is applied. Its listing is as follows:

```
{Clear the screen, and display a title}
display_title(*** General Precession ***)

{We want to use commas in the prompt string, so change the delimiter to %}
set_delimiter(%)

{What sort of coordinates do you want to precess?}
write_string(Precess ra/dec, ha/dec, az/alt, ecl or gal coordinates:)
new_line
ask_for_character(Which sort of coordinates [r,h,a,e,g]%rhaeg%c1)
```

```
{Get the coordinates...}
new_line
write_string(Please give the coordinates to be precessed:)
new_line
ask_for_coordinates(c1)

{..convert them to ra/dec if not already ra/dec...}
if (c1!r)
   new_line
   write_string(Converting to ra/dec...)
   new_line
   convert_coordinates(c1%r)
end_if

{..correct for precession..}
new_line
write_string(Correcting ra/dec for precession...)
new_line
correct_for_precession

{..convert the corrected coordinates back again..}
if (c1!r)
   new_line
   write_string(Converting from precessed ra/dec...)
   new_line
   if (c1!g)
      clear_flags(date)
   end_if
   convert_coordinates(r%c1)
end_if

{..and display the result in red}
new_line
write_string(Precessed values:)
new_line
set_normal_text(4)
display_coordinates(c1)
set_normal_text(7)

{Another conversion?}
new_line
ask_for_repeat
```

Internally, AstroScript replaces the right ascension and declination held in the data bank with the new values corrected for precession. The above AstroScript program can be used in a straightforward manner for equatorial coordinates as the following example shows.

```
              *** General Precession ***
              --------------------------

    Precess ra/dec, ha/dec, az/alt, ecl or gal coordinates:
    Which sort of coordinates [r,h,a,e,g] ........ ? R
```

```
Please give the coordinates to be precessed:
Right-ascension (h,m,s) ...................... ? 19,58,0
Declination      (d,m,s) ...................... ? 40,40,0

Correcting ra/dec for precession...
Press from date (dd.dd/mm/yyyy) ............ ? 0.9,1,1950
........ to date (dd.dd/mm/yyyy) ............. ? 28/  3/   1995
Time zone (h ahead of UT; W negative) ........ ? 0
Daylight saving (h ahead of zone t) .......... ? 0

Precessed values:
The right ascension is (h,m,s) ...............    19 59 33.8597
The declination is (d,m,s) ...............         40 47 29.1980

Again (Y or N) ............................... ? Y

Same precession epochs (Y or N) .............. ? n
Same time zone (Y or N) ...................... ? Y

                    *** General Precession ***
                    -------------------------

Press ra/dec, ha/dec, az/alt, ecl or gal coordinates:
Which sort of coordinates [r,h,a,e,g] ........ ? R

Please give the coordinates to be precessed:
Right-ascension (h,m,s) ...................... ? 19 59 33.8597
Declination      (d,m,s) ...................... ? 40 47 29.1980

Correcting ra/dec for precession...
Press from date (dd.dd/mm/yyyy) ............ ? 28/  3/   1995
........ to date (dd.dd/mm/yyyy) ............. ? 0.9,1,1950

Precessed values:
The right ascension is (h,m,s) ...............    19 58  0.0000
The declination is (d,m,s) ...................     40 39 59.9999

Again (Y or N) ............................... ? n
```

As usual, the values supplied via the keyboard have been underlined. Here, the right ascension and declination of an object at the epoch 28 March 1995 are corrected for precession to the epoch B1950.0 (just before midnight on the first of January 1950), and then corrected back again.

When using *precess.txt* for coordinates other than equatorial, you need to take care to enter the right dates. Consider the following example in which the horizon coordinates of an object are calculated for midday on 1 January 2000 from those observed at 1600 UT on 28 March 1995:

```
                    *** General Precession ***
                    -------------------------

Press ra/dec, ha/dec, az/alt, ecl or gal coordinates:
Which sort of coordinates [r,h,a,e,g] ........ ? a

Please give the coordinates to be precessed:
Azimuth  (d,m,s; N=0) ........................ ? 10,0,0          (*1)
Altitude (d,m,s) ............................. ? 30,0,0
```

```
Converting to ra/dec...
Geographic latitude (d,m,s) ................... ? 52 10 12.0000
Calendar date (d,m,y; BC neg) ................ ? 28/ 3/ 1995      (*2)
Local civil time (h,m,s) ..................... ? 16,0,0
Time zone (h ahead of UT; W negative) ........ ? 0
Daylight saving (h ahead of zone t) .......... ? 0
Geographic longitude (d,m,s) ................. ? 0 2 5.6700

Correcting ra/dec for precession...
Precess from date (dd.dd/mm/yyyy) ............ ? 28/ 3/ 1995
......... to date (dd.dd/mm/yyyy) ............ ? 1.5,1,2000

Converting from precessed ra/dec...
Calendar date (d,m,y; BC neg) ................ ? 1,1,2000           (*3)
Local civil time (h,m,s) ..................... ? 12,0,0

Precessed values:
The azimuth is (d,m,s; N=0) ..................    319 40 51.4622 (*4)
The altitude is (d,m,s) ......................     58 59 26.8457

Again (Y or N) ............................... ? Y

Same precession epochs (Y or N) .............. ? n
Same time zone (Y or N) ...................... ? Y
Same geographical longitude (Y or N) ......... ? Y
Same geographical latitude (Y or N) .......... ? Y


                    *** General Precession ***
                    -------------------------

Precess ra/dec, ha/dec, az/alt, ecl or gal coordinates:
Which sort of coordinates [r,h,a,e,g] ........ ? a

Please give the coordinates to be precessed:
Azimuth  (d,m,s; N=0) ........................ ? 319 40 51.4622    (*4)
Altitude (d,m,s) ............................. ? 58 59 26.8457

Converting to ra/dec...
Calendar date (d,m,y; BC neg) ................ ? 1/ 1/ 2000         (*3)
Local civil time (h,m,s) ..................... ? 12 0 0.0000

Correcting ra/dec for precession...
Precess from date (dd.dd/mm/yyyy) ............ ? 1.5,1,2000
......... to date (dd.dd/mm/yyyy) ............ ? 28/ 3/ 1995

Converting from precessed ra/dec...
Calendar date (d,m,y; BC neg) ................ ? 28/ 3/ 1995        (*2)
Local civil time (h,m,s) ..................... ? 16,0,0

Precessed values:
The azimuth is (d,m,s; N=0) ..................     10  0  0.0002 (*1)
The altitude is (d,m,s) ......................     29 59 59.9999

Again (Y or N) ............................... ? n
```

Here, the azimuth and altitude (*1) of an object are observed at the instant 16,0,0 UT on 28/3/1995 (*2). These coordinates are converted to right ascension and declination, and corrected for precession from 28/3/1995 to 1.5/1/2000. The precessed coordinates are then converted into their corresponding azimuth and altitude (*4) as observed at 12,0,0 UT on 1/1/2000 (*3). Next we see the process in reverse. The azimuth and altitude (*4) of the object as observed at the instant 12,0,0 UT on 1/1/2000 (*3) are converted to right ascension and declination, and corrected for precession from 1.5/1/2000 to 28/3/1995. The precessed coordinates are then converted into their corresponding azimuth and altitude (*1) as observed at 16,0,0 UT on 28/3/1995 (*2). Note the use of the *clear_flags(date)* command to make sure that we are prompted for the new dates in the middle of the program. Without it, AstroScript would use the previously entered values of date and time.

## *Nutation*

The person who has played with a spinning top may also have noticed a wobbling motion superimposed on the precession described in the previous section. This is another phenomenon common to spinning objects under the influence of external couples, and it is called *nutation*. In the case of the Earth, the nutation is excited by the gravitational couples applied by the Moon, planets, and Sun not being steady but varying as the positions of those bodies change with time. Hence the Earth's north–south axis wobbles slightly as it performs its slow precessional gyration, an effect which is usually allowed for in calculating the ecliptic longitude of a body (*nutation in longitude*), and in calculating the obliquity of the ecliptic (*nutation in obliquity*). AstroScript performs both of these corrections internally as appropriate so that you need not usually be concerned. However, there is also a single command, *compute_nutation*, which allows you to calculate and display the nutation corrections for any given moment should you so wish. For example, try the following short AstroScript program:

```
display_title(*** Nutation ***)
compute_nutation
ask_for_repeat
```

Nutation is never more than about 20 arcseconds in longitude, nor more than about 10 arcseconds in obliquity. The principal term has a period of about 18.6 years.

## *Geocentric parallax*

The apparent positions of the Moon, Sun, and other Solar-System bodies change slightly depending upon where they are observed from the surface of the Earth. The nearer the body to the Earth, the greater the apparent shift, and for the Moon this can be as much as 1 degree. In Figure 13 the Moon, M, is being observed by two observers at A and B. Each observer measures the angle of the Moon with respect to

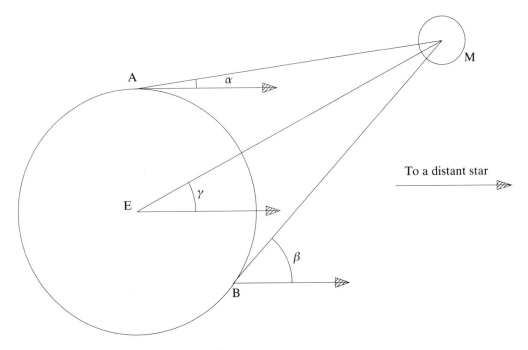

Figure 13. Geocentric parallax.

the line of sight to a very distant star. The observer at A measures the angle $\alpha$, and the one at B measures $\beta$. Clearly, $\alpha$ and $\beta$ are not the same unless M is so distant that the lines AM and BM are parallel. An imaginary observer at the centre of the Earth, E, would report yet a another value, $\gamma$. If $\gamma$ represented, say, the right ascension of the Moon as calculated from the Earth's centre (i.e. its *geocentric* position), then the observers at A and B would each have to add a different correction to $\gamma$ to obtain $\alpha$ or $\beta$, the *topocentric* positions. This shift in apparent position is called *geocentric parallax*.

When making calculations for geocentric parallax, account must be taken of the fact that the Earth is not quite a sphere, but is flattened along the polar axis, an effect probably caused by rotational forces acting on the molten Earth early in its life. The equatorial and polar radii are nevertheless quite similar, being 6378.14 km at the equator, and 6356.77 km at the poles. Positions on the Earth's surface are usually designated in terms of the *geographical* or *astronomical latitude*, $\phi$, (see Figure 14), the angle subtended at the plane of the equator by the vertical line which passes through the point on the surface and the zenith point. This is distinct from the *geocentric latitude*, $\phi'$, measured at the centre of the Earth. The two are related by

$$\rho \sin \phi' = 0.996\,647 \sin u + (h/637\,8140) \sin \phi,$$

$$\rho \cos \phi' = \cos u + (h/637\,8140) \cos \phi,$$

where

$$\tan u = 0.996\,647 \tan \phi,$$

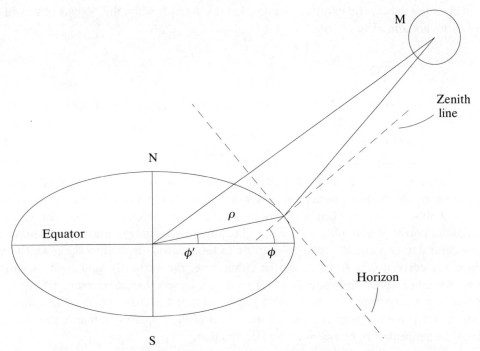

Figure 14. Allowing for the figure of the Earth. The Earth's flattened shape has been exaggerated in this diagram.

$\rho$ is the radius of the Earth at the point in question, and $h$ is the observer's height above the spheroid (sea level for all practical purposes).

Corrections for geocentric parallax may be carried out as follows. If a body has geocentric hour angle, $H$, and geocentric right ascension, $\alpha$, then its apparent coordinates, $H'$ and $\alpha'$, taking account of parallax are given by

$$H' = H + \Delta,$$

$$\alpha' = \alpha - \Delta,$$

with

$$\tan \Delta = \frac{\rho \cos \phi' \sin H}{r \cos \delta - \rho \cos \phi' \cos H'}$$

where $r$ is the distance of the body from the centre of the Earth measured in units of Earth-radii, 6378.14 km. If $r'$ is this distance in kilometres, then

$$r = \frac{r'}{6378.14}.$$

The value of $r$ can also be obtained from the *equatorial horizontal parallax* of the body,

*P*. This is the geocentric parallax measured at the equator when the body is observed to be on the horizon. Thus

$$r = \frac{1}{\sin P}.$$

The apparent declination, $\delta'$, may be obtained from the geocentric declination, $\delta$, by

$$\tan \delta' = \cos H' \frac{r \sin \delta - \rho \sin \phi'}{r \cos \delta \cos H - \rho \cos \phi'}.$$

AstroScript contains algorithms for performing all these corrections. They are usually made internally, without bothering you with the details. Occasionally, however, you might wish to make explicit corrections yourself, in which case you can use the command *correct_for_parallax(⟨string⟩)*. The argument ⟨*string*⟩ must be set either to 'topocentric to geocentric', or to 'geocentric to topocentric' depending upon which way round the correction is to be made. In either case, the currently held right ascension and declination (previously supplied, for example, by *ask_for_coordinates(r)*) are used as starting values and are replaced by the corrected values. Hence the following AstroScript program fragment can be used to find the apparent azimuth and altitude given the geocentric right ascension and declination:

```
{Get the geocentric right ascension and declination}
ask_for_coordinates(r)

{Correct for geocentric parallax}
correct_for_parallax(geocentric to topocentric)

{Convert to azimuth and altitude}
convert_coordinates(r,a)

{Display the result}
display_coordinates(a)
```

You can embellish this to perform the correction either way around by using the *ask_for_character* command and an *if* command to set the argument string s1 either to 'geocentric to topocentric', or 'topocentric to geocentric' as appropriate, e.g.:

```
{Which way round should we do it?}
ask_for_character(Convert from geo or topocentric coords [g/t],gt,c1)

{Set s1 as appropriate}
if (c1=g)
    set_string(s1,geocentric to topocentric)
end_if

if (c1=t)
    set_string(s1,topocentric to geocentric)
end_if
```

{Now perform the conversion}
correct_for_parallax(s1)

Further embellishments could include generalising the type of coordinates to be corrected etc.

## *Atmospheric refraction*

The apparent position of a celestial body viewed from the surface of the Earth is affected by the presence of the atmosphere. This is because the refractive index of the atmosphere is not quite unity, and its curvature (following that of the Earth) makes it act like a weak lens causing the path of a ray to be diverted from that which it would have followed had there been no atmosphere at all. This effect is called *atmospheric refraction*. The situation is illustrated in Figure 15. The line marked A is the path of a ray of light coming towards a point on the Earth's surface which it would follow if there were no atmosphere. It is of course a straight line. The path marked B is the path actually followed by a ray coming from the same direction as A, but which is bent by the atmosphere to arrive at the same point on the Earth. The angle marked $\zeta$ is the apparent zenith direction of the ray B, whilst $z$ is the zenith direction with no atmosphere. The angles $z$ and $\zeta$ are different from one another (except at the zenith itself where both are zero). The difference increases with zenith angle, reaching a

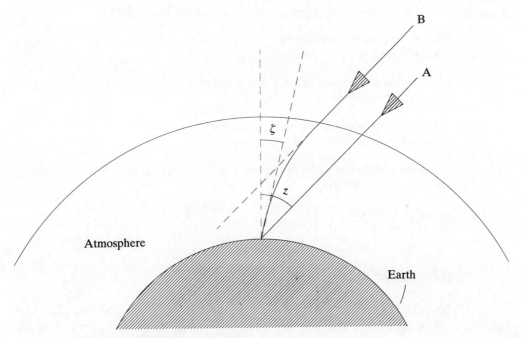

Figure 15.  Atmospheric refraction.

maximum of about 34 arcminutes at the horizon. The exact value at any altitude depends upon the meteorological conditions, but there exist standard formulae which give average values dependent only on the temperature and pressure at the place of observation. If we observe a star with apparent zenith angle $\zeta$, its true zenith angle, $z$, is given by

$$z = \zeta + R,$$

where for altitudes greater than 15 degrees

$$R = \frac{0.004\,52\,P\tan z}{273 + T}\ \text{degrees},$$

or for altitudes below 15 degrees

$$R = \frac{P(0.1594 + 0.0196a + 0.000\,02a^2)}{(273 + T)(1 + 0.505a + 0.0845a^2)}\ \text{degrees},$$

with $P$ being the barometric pressure in millibars, $T$ the temperature in degrees centigrade, and $a$ the apparent altitude in degrees.

AstroScript has algorithms for making such corrections which can be invoked using the command *correct_for_refraction(⟨string⟩)*. The argument ⟨*string*⟩ can be set to 'true to apparent' or 'apparent to true' depending on the direction in which the correction is to be made. The currently held values of the right ascension and declination (supplied for example using the *ask_for_coordinates(r)* command) are converted and replaced with the corrected values. In the following AstroScript program fragment, you are asked to supply the apparent horizon coordinates (azimuth and altitude) of a celestial object and the program calculates the true values:

```
{Get the apparent coordinates}
ask_for_coordinates(a)

{Convert them to right ascension and declination}
convert_coordinates(a,r)

{Replace them with the values corrected for refraction}
correct_for_refraction(apparent to true)

{Convert back to horizon coordinates}
convert_coordinates(r,a)

{Display the result}
display_coordinates(a)
```

## Aberration

Imagine that you are standing still in the rain on a windless day. The rain falls vertically downwards, and you can protect yourself by holding your umbrella straight above your head. Now imagine that you are on a bicycle riding through the rain on the same windless day. Because of your forward motion, you now see the rain coming down towards you at an angle from the vertical direction, and the faster you travel the greater the angle. You now have to slant the umbrella forward to keep the rain off your face. Exactly the same thing happens with light. Imagine that you are viewing a star at the zenith on the surface of the Earth which has, for the purposes of this experiment, stopped dead in its orbit around the Sun. The light rays from the star come vertically downwards, and you have to set your telescope straight up to view the star. Now let the Earth continue in its orbit again. It moves at about $30 \, \text{km} \, \text{s}^{-1}$, and light travels at about $300\,000 \, \text{km} \, \text{s}^{-1}$. The result is that the light rays from the star appear now to come from a small angle to the vertical in the direction of the Earth's motion, and you have to tilt your telescope slightly to see the star. This is called *aberration*, and it affects the apparent positions of all celestial objects, not just those at the zenith. It is never more than 20.5 arcseconds in magnitude.

The effect of aberration on the ecliptic longitude and latitude, and on the right ascension and declination can easily be calculated. If $(\lambda, \beta)$ and $(\alpha, \delta)$ are the ecliptic and equatorial coordinates of a celestial body, then the corrections $(\Delta\lambda, \Delta\beta)$ and $(\Delta\alpha, \Delta\delta)$ which must be added to correct for aberration are given by

$$\Delta\lambda = \frac{-20.495\,52 \cos(\lambda_s - \lambda)}{\cos\beta} \text{ arcseconds,}$$

$$\Delta\beta = -20.495\,52 \sin(\lambda_s - \lambda)\sin\beta \text{ arcseconds,}$$

$$\Delta\alpha = -20.495\,52 \frac{\cos\alpha \cos\lambda_s \cos\varepsilon + \sin\alpha \sin\lambda_s}{\cos\delta} \text{ arcseconds,}$$

and

$$\Delta\delta = -20.495\,52[\cos\lambda_s \cos\varepsilon(\tan\varepsilon\cos\delta - \sin\alpha\sin\delta) + \cos\alpha\sin\delta\sin\lambda_s]$$
$$\text{arcseconds,}$$

where $\lambda_s$ is the ecliptic longitude of the Sun and $\varepsilon$ is the obliquity of the ecliptic. The apparent coordinates, $(\lambda', \beta')$ and $(\alpha', \delta')$, are then given by

$$\lambda' = \lambda + \Delta\lambda,$$

$$\beta' = \beta + \Delta\beta,$$

$$\alpha' = \alpha + \Delta\alpha,$$

$$\delta' = \delta + \Delta\delta.$$

The above expressions make the assumption that the Earth's orbit is circular, and ignore small additional corrections arising as a result of the slightly elliptical orbit. These never amount to more than about 0.4 arcseconds.

AstroScript makes corrections for aberration when calculating the apparent positions of some Solar-System bodies, but does not support a general correction command in the current version. The positions of all bodies outside the Solar System in a given direction are similarly affected, so aberration only produces relative shifts between objects lying far apart on the sky.

## *Rising and setting*

The rotation of the Earth about its north–south axis makes it seem as if the objects in the sky are in constant motion. We can imagine that there is a point in the sky (the celestial pole) where the axis cuts the celestial sphere, and all the stars appear to move in circles around it making one complete revolution in 24 sidereal hours. At present, a star called Polaris is quite close to the north celestial pole, so it seems as if everything revolves around it, but precession will cause it gradually to move away during the next few thousand years until it no longer occupies that central position. There is no corresponding star at the south celestial pole.

The altitude of the celestial pole above the horizon depends on an observer's position on the Earth. At the geographical north pole, the north celestial pole is right overhead in the zenith, whilst at the geographical equator both north and south celestial poles appear to be on the horizon. A star with a small enough angular distance from the celestial pole stays above the horizon all day, and is said to be *circumpolar*. It crosses the observer's meridian twice per day, at *upper transit* or *upper culmination* when at greatest altitude, and *lower transit* or *lower culmination* when at smallest altitude. Stars with larger angular distances from the celestial pole dip lower in the sky, and if the angular distance is large enough the star disappears below the horizon for part of the day. As it crosses the horizon on the way down it is said to be *setting*, and as it crosses the horizon on the way up it is said to be *rising*. Stars further away still from the celestial pole may never rise above the horizon at all but stay permanently out of sight.

The times and azimuths of rising and setting can be calculated by the formulae

$$\mathrm{LST_r} = 24 + \alpha - H,$$

$$\mathrm{LST_s} = \alpha + H,$$

$$\cos H = -\frac{\sin d + \sin \phi \sin \delta}{\cos \phi \cos \delta},$$

$$\cos A_r = \frac{\sin \delta + \sin d \sin \phi}{\cos d \cos \phi},$$

$$A_s = 360 - A_r,$$

where $\mathrm{LST_r}$ and $\mathrm{LST_s}$ are the local sidereal times of rising and setting respectively, $\alpha$ is the right ascension of the object, $H$ is the hour angle, $A_r$ and $A_s$ are the azimuths of rising and setting, respectively, $\delta$ is the declination of the object, $\phi$ is the geographical latitude, and $d$ is the *vertical displacement* at the horizon. LST, $\alpha$, and $H$ are all measured in hours; $A$, $\delta$, and $d$ are all measured in degrees. The parameter $d$ takes account of the facts that (i) your horizon may have hills on it or that you are not at sea level yourself, (ii) the celestial object under consideration may have an appreciable angular size (like the Sun) so that the coordinates of its centre are not the same as those of its upper limb, and (iii) atmospheric refraction causes a displacement upwards of about half a degree in the apparent position at the horizon. The value of *d* is zero for a point object viewed over the sea from sea level with no atmosphere. Making it more positive delays the time of setting and brings forward the time of rising. The standard correction for atmospheric refraction is +34 arcminutes and the quarter-degree angular radii of the Sun and Moon add a further 15 arcminutes to *d* for the upper limb.

You can make rising and setting calculations very easily with AstroScript. There are two commands, *find_rise_set(⟨string⟩)* and *display_rise_set(⟨string⟩)*, which take care of all the details for you. The argument ⟨*string*⟩ can be set to any of 'sun', 'moon', 'mercury', 'venus', 'mars', 'jupiter', 'saturn', 'uranus', or 'neptune' to find the circumstances of rising and setting for the named object. In each case, the computation has to proceed through several iterations since the coordinates of the object are changing constantly as it moves in its orbit, so that its position at, say, setting is unknown until the time of setting has been found. For example, you could use the following AstroScript lines to calculate and display the times of moonrise and moonset:

```
find_rise_set(moon)
display_rise_set(moon)
```

The argument ⟨*string*⟩ may also be set to any other name, such as 'Ceres' or 'Arcturus', or 'Fred', or indeed left blank (write as *find_rise_set()*). In this case, AstroScript assumes that you wish to make the calculation for the object whose position is given by the current values of right ascension and declination (perhaps previously entered using *ask_for_coordinates(r)*). For example, these lines allow you to find the circumstances of rising and setting for any given object whose coordinates you supply:

```
ask_for_coordinates(r)
find_rise_set()
display_rise_set()
```

A more-general program is supplied on the disk which comes with the book under the name of *rise_set.txt*. Here is its listing:

```
{Clear the screen and put up a title}
display_title(+++ Rising and Setting +++)

{Set the delimiter to % so that we can use commas in prompts}
set_delimiter(%)

{Set the default object name}
if (s1=.)
   set_string(s1%Sun)
end_if

{Get the name of the object}
ask_for_string(Object name [Sun, Moon, a planet etc.]%s1)
```

```
{Get the ra & dec if necessary}
if (s1!sun)
  if (s1!moon)
    if (s1!mercury)
      if (s1!venus)
        if (s1!mars)
          if (s1!jupiter)
            if (s1!saturn)
              if (s1!uranus)
                if (s1!neptune)

{Phew! none of those, so get ra/dec}
                ask_for_coordinates(r)
              end_if
            end_if
          end_if
        end_if
      end_if
    end_if
  end_if
end_if
end_if

{Calculate the rising and setting...}
find_rise_set(s1)

{...and display the result in a good colour}
set_normal_text(10)
display_rise_set(s1)
set_normal_text(7)

{Go again?}
ask_for_repeat
```

Note that *find_rise_set* and *display_rise_set* calculate and display the circumstances of rising and setting of the given object on the specified calendar date. The time of setting of any object may well be earlier than that of rising, except in the case of the Sun for which setting always follows rising. In making the calculation, *find_rise_set* takes account of the vertical displacement at the horizon, $d$, using both the value of *vertical_shift* supplied by the setup.dat file or specified by the *set_vertical_shift* command, and the height above mean sea level supplied in the course of the computation. You can take account of atmospheric refraction, the finite angular size of an object's disk, and distant hills raising your horizon above mean sea level with *vertical_shift*. The effect of your own height above sea level is automatically taken into account. Thus, if atmospheric refraction contributed 34 arcminutes and your horizon was half a degree higher than mean sea level, you would need to set *vertical_shift* to $34 - 30 = 4$ arcminutes.

# *Orbits*

The motions of the celestial bodies in our Solar System are all controlled by the action of gravity, i.e. that mysterious universal force of attraction experienced between any two masses. The force, $F$, is proportional to the product of the masses, $m$ and $M$, of the bodies involved, and inversely proportional to the square of the separation, $r$, between their centres of mass:

$$F = G\frac{mM}{r^2},$$

where $G$ is the universal gravitational constant $(6.672 \times 10^{-11}\,\mathrm{N\,m^2\,kg^{-2}})$. One of the consequences of the inverse square law was pointed out by Isaac Newton. He showed that stable bound orbits were possible such that the orbiting body described an exactly repeating ellipse about the central mass. This is more or less the situation in the Solar System. The mass of the Sun $(1.9891 \times 10^{30}\,\mathrm{kg})$ is so large compared with all the other bodies that its gravitational field dominates nearly everywhere except very close to major planets, moons, and asteroids. Hence the planets describe nearly perfect elliptical orbits about the Sun, being perturbed only slightly by the gravitational influences of the other members of the Solar System.

In the general case, the orbit followed by a body of mass $m$ about a very massive body of mass $M$ depends on its total energy, $E$. In Figure 16, P marks the position of the orbiting body at distance $r$ from the attracting centre S. The body is travelling at speed $v$ in its orbit. Point A marks the point of closest approach between the bodies. The total energy, $E$, has a contribution from the *kinetic energy*, $mv^2/2$, of the small body (its energy as a result of being in motion), and the *potential energy*, $-GmM/r$, which is a measure of how much kinetic energy could be gained by the orbiting body were it to fall in towards the attracting centre. Thus

$$E = \frac{mv^2}{2} - G\frac{mM}{r}.$$

Whether the orbit is *closed* and *bound*, or *open* and *unbound*, depends on whether $E$ is less than or greater than zero (see Figure 17). If $E$ is less than zero, the kinetic energy is insufficient for the orbiting body to escape from the mutual gravitational attraction. It therefore remains continuously within the gravitational field and, since the force of attraction varies as the inverse square of the separation, the orbit is closed and elliptical in shape. The circular orbit (Figure 17(a)) is a special case of an ellipse where the central mass is at the centre of the circle. In the general case (Figure 17(b)) the

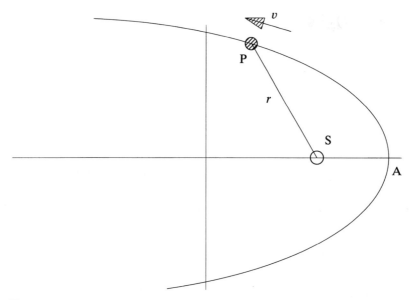

Figure 16.  A general orbit.

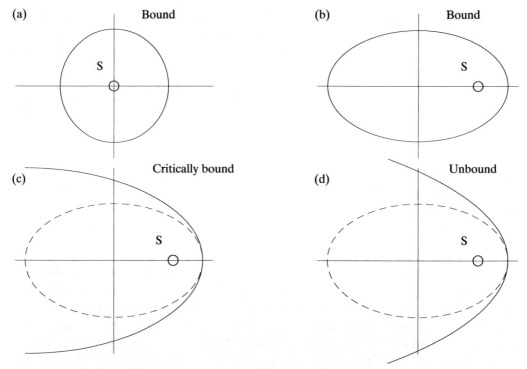

Figure 17.  Bound and unbound orbits.

central mass (the Sun in our Solar System) is at one of the two *foci* of the ellipse. As the kinetic energy is increased, the ellipse gets larger and larger, until the body has just enough energy to escape from the gravitational field at infinite distance. The value of $E$ is zero in this case; the orbit is on the borderline between being bound and unbound (Figure 17(c)), and follows a *parabolic* curve. With yet more kinetic energy, (Figure 17(d)) the gravitational field is no longer strong enough to bind the orbiting body to the central mass. Even at infinite distance, the body still has some kinetic energy, $E$ is greater than zero, the orbit is unbound, and it follows a *hyperbolic* curve.

## Elliptical orbits

An orbital ellipse is illustrated in Figure 18. You can imagine the ellipse as a squashed circle. The two foci are S and S′, with the attracting centre (e.g. the Sun) at S. The amount of squashing is measured by the *eccentricity*, $e$, of the ellipse. The eccentricity is zero for a circle and S and S′ are then coincident in the centre. The most flattened ellipses have eccentricities approaching 1, and S and S′ are almost at points A and B respectively. The size of the ellipse is determined by the length of the *semi-major axis*, $a$. The distance of each focus from the centre is then $ea$, and the length of the *semi-minor axis*, $b$, is given by

$$b^2 = a^2(1 - e^2).$$

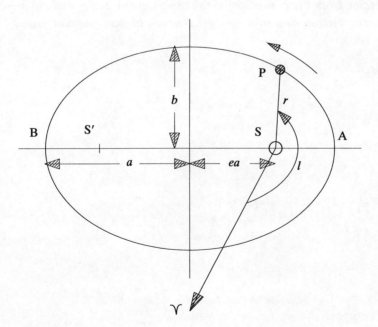

Figure 18. An orbital ellipse.

Most planetary orbits have eccentricities less than 0.1 so that deviations from true circular motion are often quite small.

Imagine in Figure 18 that P marks the position of a planet in orbit about the Sun, S. If that planet is the Earth, then the plane of the ellipse defines the *plane of the ecliptic*. The intersection between it and the (extended) plane of the Earth's equator defines the primary direction, *the vernal equinox* or *first point of Aries*, from which all orbital positions are measured. This is marked ♈ in the diagram. Thus ♈ lies in the Earth's orbital plane, but does not do so for the other planets since their orbits are all inclined at small angles to the plane of the ecliptic. The position of the planet P in its orbit is measured by its heliocentric *orbital longitude*, $l$, which is the angle at the Sun between the vernal equinox and the *radius vector*, $r$, computed as the sum of angles in two different planes except in the case of the Earth.

The general case for a planet is illustrated in Figure 19. The Sun, S, is at the centre of the diagram, and you are to imagine that you are looking at the path of the planet, P, around the Sun from a great distance. The orbit of the planet is the ellipse $N_1APN_2B$, and that part lying above (north of) the ecliptic is shown by the solid curve $N_1APN_2$ whilst that lying below it is marked by the dashed curve $N_2BN_1$. Point A is the position of closest approach to the Sun, the *perihelion*, and point B is the position of greatest distance from the Sun, the *aphelion*. The line $N_1N_2$ is the line of intersection of the planet's orbital plane with the ecliptic. The point $N_1$ is where the planet rises above the ecliptic from below, and is called the *ascending node*; $N_2$ marks the point where the planet moves below the ecliptic from above, and is called the *descending node*.

The planet's position in its orbit is specified by the *true anomaly*, $v$, measured in the direction of motion from the perihelion. If the orbit were circular, $v$ would increase at a constant rate, completing one orbit of 360 degrees in one periodic time. The true

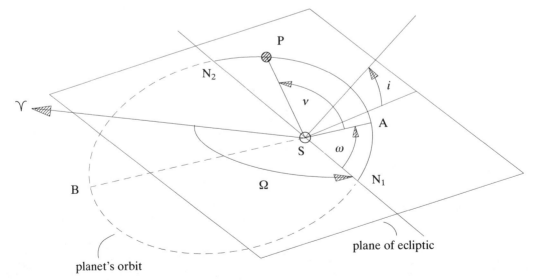

Figure 19. The orbit of a planet around the Sun.

anomaly increases at a variable rate in an elliptical orbit, the rate being greatest at the perihelion, and least at the aphelion. We can imagine a fictitious body, P′, moving in a circular orbit at constant speed and making one complete revolution in the same period as the actual planet, P. Both start together at A and coincide again at B. The angle made with the major axis by this 'mean' planet is called the *mean anomaly*, $M_a$, (see Figure 20) and its relationship with the true anomaly is sometimes known as the *equation of the centre*:

$$v = M_a + 2e \sin M_a.$$

This is actually an approximate solution of *Kepler's equation*, which relates the true anomaly to the mean anomaly using an intermediate quantity called the *eccentric anomaly*, $E$. Kepler's equation is

$$E - e \sin E = M_a,$$

from which the true anomaly may be found using

$$\tan \frac{v}{2} = \sqrt{\frac{1+e}{1-e}} \tan \frac{E}{2}.$$

Kepler's equation cannot be solved directly but can be tackled using an iterative procedure (see, for example, *Practical Astronomy with your Calculator*, third edition, page 90; see *Bibliography*).

The orientation of the planet's orbital ellipse with respect to the ecliptic is described by three parameters. The *longitude of the ascending node*, $\Omega$, is measured in the plane of the ecliptic from the direction of the vernal equinox in the same sense as the motion of all the planets around the Sun. (It is a curious fact, and one which is a clue to the origin of the Solar System, that all the planets progress in the same sense, anticlockwise when viewed from above (north of) the ecliptic. Any rotation in this sense is termed *prograde motion*, and the opposite sense *retrograde motion*.) The position of the perihelion, A, is described by the angle at the Sun between it and the ascending node, and is measured in the plane of the orbit. This is called the *argument of the perihelion*, $\omega$. Often, $\omega$ is added to $\Omega$ to give the *longitude of the perihelion*, $\varpi = \Omega + \omega$. Note that this is the sum of two angles in different planes. The third parameter is the *inclination* of the orbit, $i$, which is just the angle between the planes of the orbit and the ecliptic.

The heliocentric longitudes of the 'true' planet and the 'mean' planet can now be specified. Both quantities are measured from the direction of the vernal equinox, and involve angles in two different planes (except for the Earth whose orbit is in the plane of the ecliptic). The mean longitude of the planet is just $M_a + \omega + \Omega$, and the true longitude is $v + \omega + \Omega$.

It is clear from the foregoing discussion that you need several parameters to describe the position of an orbiting body at any moment. These include $\Omega$, $\omega$, and $i$ which describe the orientation of the orbital ellipse, and $v$ which fixes the planet's position in its orbit. Sometimes $v$ is not given explicitly, but must be calculated from other

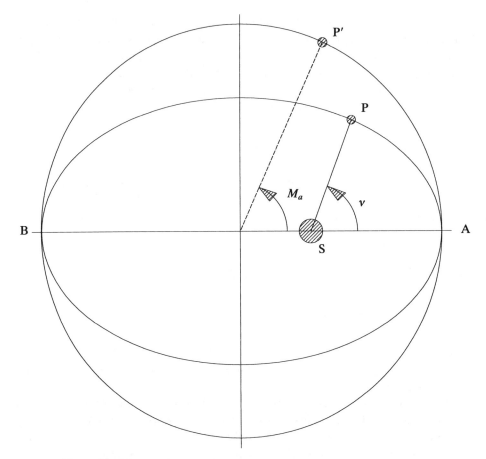

Figure 20. Mean and true anomalies.

parameters, such as the *period*, *T*, (usually in years) and the *longitude at the epoch*, i.e. the true longitude of the planet at a given moment. Two other parameters are needed to complete the list which specify the size and shape of the ellipse itself. They are the *eccentricity*, *e*, and the length of the *semi-major axis*, *a*. Given a complete set of parameters, you can calculate the position of the orbiting body exactly for any moment in the future or the past, provided of course that the orbit is indeed a true ellipse. Whilst quite good results are obtained on this assumption, the effects of the perturbing influences of the other planets need to be taken into account for high precision. They cause the 'constant' parameters of the orbit not to be constant any more; for example, the longitude of the ascending node may change slowly with time.

## *Low-precision orbit calculations*

AstroScript makes accurate calculations of orbital motion taking account of perturbations where possible using the method outlined in the next section. However, it may be helpful first to run through an example calculation of true elliptical motion, which is almost always an approximation to the real motion. Let us find the position of Venus at $0^h$ on 21st April, 1984 on the assumption that its orbit is an unperturbed ellipse. We start with the elements of the Venusian orbit as follows:

*Elements of the orbit of Venus at epoch 1990 January 0.0:*

| | | |
|---|---|---|
| period: | 0.615 211 | tropical years |
| longitude at the epoch: | 88.455 855 | degrees |
| longitude of perihelion: | 131.430 236 | degrees |
| eccentricity: | 0.006 778 | |
| semi-major axis: | 0.723 332 | au |
| inclination: | 3.394 535 | degrees |
| longitude of ascending node: | 76.589 820 | degrees |

Our calculation will proceed in three steps. The first is to find the position of Venus in its own orbital plane, i.e. find the true anomaly. Next, we project the planet's calculated position onto the plane of the ecliptic to find its *heliocentric* ecliptic longitude and latitude. Finally, we transform from the Sun to the Earth to find the equivalent *geocentric* ecliptic coordinates.

The mean anomaly of Venus at $0^h$ on 21st April, 1984, can be found from

$$M_a = \frac{360}{365.242\,191} \cdot \frac{D}{T} + \varepsilon - \varpi \text{ degrees,}$$

where $T$ is the orbital period in tropical years, $\varepsilon$ is the longitude at the epoch, $\varpi$ is the longitude of the perihelion, and $D$ is the number of days (including the fraction) since the epoch. The value of $D$ is best calculated using Julian day numbers, i.e.

$$D = JD - JD_0,$$

where $JD$ is the Julian day number of $0^h$ on 21st April, 1984, and $JD_0$ is the Julian day number of the epoch. We can use the AstroScript program *calendar.txt* to find $JD$ and $JD_0$, and hence $D = 2\,445\,811.5 - 2\,447\,891.5 = -2080.0$. Then $M_a = -3375.402\,606$ degrees, which we can reduce to the primary range of 0–360 by adding multiples of 360: $M_a = 224.597\,394$ degrees. The equation of the centre then gives us

$$v = M_a + (360e \sin M_a)/\pi = 224.052\,056 \text{ degrees.}$$

The planet's true orbital longitude is therefore

$$l = v + \varpi = 355.482\,292 \text{ degrees.}$$

We also need to calculate the radius vector which is given by

$$r = \frac{a(1 - e^2)}{1 + e \cos v} = 0.726\,839 \text{ au,}$$

where $a$ is the length of the semi-major axis, and the heliocentric latitude of the planet, $\psi$,

$$\psi = \arcsin\left[\sin\left(l - \Omega\right)\sin i\right] = -3.353\,686 \text{ degrees},$$

where $i$ is the inclination of the orbit.

Now we need to project our calculations onto the plane of the ecliptic to obtain the projected heliocentric longitude, $l'$, and the projected radius vector, $r'$, as follows:

$$l' = \arctan\left[\tan\left(l - \Omega\right)\cos i\right] + \Omega = 355.497\,671 \text{ degrees},$$

$$r' = r\cos\psi = 0.725\,594 \text{ au}.$$

The final step in the process is to refer the calculation to the Earth to find the geocentric ecliptic longitude, $\lambda$, and latitude, $\beta$, of the planet. For this we must know the Earth's true longitude, $L$, and radius vector, $R$, which we can calculate using the same formulae as for $l$ and $r$ above. The Earth's orbital elements are:

*Elements of the orbit of Earth at epoch 1990 January 0.0:*

| | | |
|---|---|---|
| *period:* | *1.000 040* | *tropical years* |
| *longitude at the epoch:* | *99.403 308* | *degrees* |
| *longitude of perihelion:* | *102.768 413* | *degrees* |
| *eccentricity:* | *0.016 713* | |
| *semi-major axis:* | *1.000 000* | *au* |

The result for $0^h$ on 21st April, 1984 is $L = 211.174\,442$ degrees and $R = 1.005\,024$ au. The formula for finding the geocentric ecliptic longitude depends on whether the planet is an inner planet ($r' < R$; Mercury and Venus) or an outer planet ($r' > R$; Mars, Jupiter, Saturn, Uranus, Neptune, and Pluto). For the former case we have

$$\tan\left(\lambda - l'\right) = \frac{R\sin\left(l' - L\right)}{r' - R\cos\left(l' - L\right)} \quad \text{(inner planet)},$$

and for the latter

$$\tan\left(\lambda - 180 - L\right) = \frac{r'\sin\left(L - l'\right)}{R - r'\cos\left(L - l'\right)} \quad \text{(outer planet)}.$$

The formula for the geocentric latitude is the same for inner and outer planets alike:

$$\tan\beta = \frac{r'\tan\psi\sin\left(\lambda - l'\right)}{R\sin\left(l' - L\right)}.$$

Substituting values we find $\lambda = 16.360\,912$ degrees, and $\beta = -1.479\,899$ degrees. The more-accurate routines in AstroScript which take account of perturbations give $\lambda = 16.225\,642$ degrees and $\beta = -1.476\,462$ degrees, so our simple method has reached within about 0.14 degrees in longitude, and 0.004 degrees in latitude. Some of the error is caused by the perturbations to the orbit of Venus, ignored in our simple method, but there is also error in using only the first term in the equation of the centre rather than solving Kepler's equation properly, and in ignoring the effect of light time. You can make more-exact elliptical-orbit calculations by specifying the argument string 'elliptical' in the AstroScript command *find_position* (see later).

## *Perturbations and osculating elements*

High-precision orbital calculations require that the perturbing influences of the other members of the Solar System are allowed for. These cause constantly changing slight deviations from true elliptical motion. One way to take account of perturbations in orbital calculations is to expand an object's heliocentric longitude, latitude, and radius vector in terms of sinusoidal components (sines and cosines), to produce a so-called *theory*. The arguments of the sinusoids denote the frequencies of the perturbing terms, and their amplitudes decrease quickly as the frequency increases, so that a realistic level of accuracy can be achieved by truncating the series after relatively few terms. This is the approach used by Meuus, for example, in his excellent *Astronomical Algorithms* (see *Bibliography*), and is adopted here in AstroScript. Planetary calculations are based on the theory VSOP 87 by Bretagnon & Francou, and lunar calculations are based on the theory ELP 2000 by Chapront-Touze & Chapront (see *Bibliography*).

The true heliocentric longitude, $l$, latitude, $b$, and radius vector, $r$, are each represented by a constant term plus five contributions which depend on the number of Julian millennia, $\tau$, of 365 250 days since the epoch J2000 as follows:

$$l = l_0 + l_1\tau + l_2\tau^2 + l_3\tau^3 + l_4\tau^4 + l_5\tau^5,$$

$$b = b_0 + b_1\tau + b_2\tau^2 + b_3\tau^3 + b_4\tau^4 + b_5\tau^5,$$

$$r = r_0 + r_1\tau + r_2\tau^2 + r_3\tau^3 + r_4\tau^4 + r_5\tau^5.$$

Each of these contributions, $l_1$, $l_2$, etc. represents a sum of other terms of the general form $A\cos(B + C\tau)$, where $A$, $B$, and $C$ are constants. In the case of Venus, for example, $r_1$ is given by

$$r_1 = 34\,551\cos(0.891\,99 + 10\,213.285\,55\tau) + 234[\cos(1.772 + 20\,426.571\tau) - 1.0],$$

in units of $10^{-8}$ arcseconds. AstroScript incorporates many hundreds of such terms, but you need not be concerned with the details. It is not particularly instructive to try to identify each term as caused by the influence of a particular body. Rather, we should recognise that calculations of many-body interactions are intractable, and the complexity of the method outlined here just reflects this. Even after taking account of hundreds of terms, the result is still an approximation – but quite a good one!

Many objects in the Solar System have either not had theories developed for them, or their motions are so irregular as to make it impractical to describe their orbits as a series of terms. An example is the orbit of Pluto, which weaves in and out of that of Neptune and hence is heavily perturbed by the gravitational influences of that planet. In these cases, it is often convenient to describe their orbits over restricted intervals of time using *osculating elements*. The idea is that the motion approximates very closely to an ellipse for a short period, so can be described by the elements for that ellipse without the need for perturbing terms. Of course, the true motion gradually departs from the ideal, and soon it is necessary to use a new set of osculating elements. Arbitrarily high precision can be achieved within the interval for which the elements are valid, however, with relatively simple calculations of the sort we made above for Venus. AstroScript provides support for osculating calculations too (see later).

## *Parabolic orbits*

Some members of the Solar System, especially the comets, come from such a large distance from the Sun that they have almost escaped from the solar gravitational field. It makes sense therefore to describe their orbits near the Sun in terms of parabolic rather than elliptical trajectories. The orbits are usually heavily perturbed as well, so that osculating parabolic elements are needed. The parabolic elements are similar to those for elliptical orbits, except that there is no orbital period (an object travelling on a truly parabolic trajectory never returns), the eccentricity has no meaning, and the 'size' of the orbit is described in terms of the distance of closest approach to the Sun (the perihelion distance). In addition, you usually need to supply the epoch at which the comet is at perihelion. You can make parabolic calculations in AstroScript (see below).

## *Calculating an orbital position*

The first step in calculating an orbital position, whether it be a planet, or another member of the Solar System, and whether the orbit is described in terms of a theory or by osculating elements, is to find the heliocentric longitude, $l$, latitude, $b$, and radius vector, $r$. The same quantities must also be found for the Earth, designated as $L$, $B$, and $R$. The value of $B$ is always within 0.4 arcseconds of zero. We can use the method outlined earlier in this chapter to find the geocentric ecliptic coordinates, $(\lambda, \beta)$, or we may use

$$\tan \lambda = \frac{y}{x},$$

$$\tan \beta = \frac{z}{\sqrt{x^2 + y^2}},$$

where

$$x = r \cos b \cos l - R \cos B \cos L,$$

$$y = r \cos b \sin l - R \cos B \sin L,$$

and

$$z = r \sin b - R \sin B.$$

These coordinates may then be transformed into any other system using the procedures outlined earlier in this book (e.g. *convert_coordinates*).

A complication arises when the orbiting body is a long way from the Earth. For high precision, it is then necessary to include the effects of *light time*. Light from the body may take many hours to reach the Earth, so that the position we observe now is that of the object several hours ago. The usual way to allow for this is to make the calculation twice. On the first pass, you calculate the Earth–body distance, $r$, for the prescribed dynamic time, say $t_0$. On the second pass, you make the calculations for the moment $t_0 - r/c$, where $c$ is the speed of light ($299\,792\,458 \text{ m s}^{-1}$). You might also need to take account of aberration, nutation, and geocentric parallax.

You can make accurate orbital calculations easily using AstroScript commands (see below).

## The phase, position angle, parallactic angle, and vertical angle of the bright limb

At any point in the orbit of a planet or the Moon, the hemisphere which faces towards the Sun is brightly illuminated whereas the other half is dark. The fraction of the surface which we can see from the Earth, however, usually overlaps both bright and dark sides, so we are presented with a view of the disk which is not uniformly illuminated but which contains a bright segment, the rest being dark and often invisible. As the relative positions of the Earth, the object, and the Sun vary, so the area of the disk that is illuminated changes. The *phase* is defined to be the fraction of the disk that is illuminated.

Figure 21 is a diagram of the appearance of a planet or the Moon when it is not uniformly illuminated. The curve dividing the bright side from the dark side is called the *terminator*, and is elliptical in shape, meeting the edge of the disk at points a and b. The line m is perpendicular to the straight line ab through the centre of the disk, and points towards the Sun. The straight line NS marks the trace of the Earth's north–south axis projected onto the disk, and the line Z marks the trace of the observer's vertical direction.

The phase, $F$, of the planet or Moon is given by the expression

$$F = (1.0 + \cos d)/2,$$

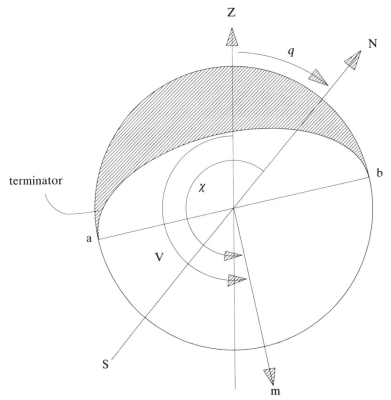

Figure 21. The bright limb of a planet or the Moon.

with $d = l - \lambda$ for a planet, where $l$ is the true heliocentric longitude and $\lambda$ is the true geocentric longitude. The expression for $d$ in the case of the Moon is a little more complicated:

$$\cos \varphi = \cos \beta \cos (\lambda - \lambda_s),$$

$$\tan d = \frac{R \sin \varphi}{\Delta - R \cos \varphi},$$

where $(\lambda, \beta)$ are the geocentric ecliptic longitude and latitude of the Moon, $\lambda_s$ is the geocentric ecliptic longitude of the Sun, $R$ is the Earth–Sun distance, and $\Delta$ is the Earth–Moon distance.

The orientation, or *position angle*, of the bright limb is determined by the direction of the line m measured anticlockwise from the north point, represented by $\chi$ in Figure 21. The angle, V, of line m measured from the vertical direction is the *vertical angle* of the bright limb, and is the angle of orientation seen by an observer from the ground. The angle between Z and N is called the *parallactic angle*, $q$, which varies with hour angle, being zero when the object crosses the meridian in upper culmination. It is caused by the fact that the object's motion on the sky follows the curve of a small circle, or *parallel* circle, and has nothing to do with parallax. The parallactic angle can be calculated using the formula

$$\tan q = \frac{\sin H}{\tan \phi \cos \delta - \sin \delta \cos H},$$

where $H$ is the hour angle, $\phi$ is the geographical latitude, and $\delta$ is the declination.

## *Orbital calculations with AstroScript*

Finding the position of an orbiting body is especially easy with AstroScript. There are two primary commands, *find_position(⟨string⟩)* and *display_position (⟨string⟩)*. The complexities of the foregoing discussion are kept strictly out of sight, and all you need to do is to set the argument ⟨*string*⟩ to specify what sort of calculation you wish to perform. Valid settings are 'sun', 'moon', 'mercury', 'venus', 'mars', 'jupiter', 'saturn', 'uranus', 'neptune', 'elliptical', and 'parabolic'. The last two in the list, 'elliptical' and 'parabolic', refer to osculating orbit calculations in which you are asked to supply the elliptical or parabolic elements of the orbit, respectively. Thus to find and display the position of the Sun, you could write

    find_position(sun)
    display_position(sun)

and that's all. AstroScript would make the calculation of the Sun's position using the full perturbation theory outlined above and then display the result. Try it now for yourself.

The results displayed by the *display_position* command depend slightly on the particular class of object being displayed (see Table 1). For the Sun *true* values of the ecliptic longitude, ecliptic latitude, Sun–Earth radius vector, right ascension, and declination are displayed which have not been corrected for parallax, nutation, or aberration. *Apparent* values of the ecliptic longitude, right ascension, and declination include corrections for nutation and aberration, and *topocentric* values of the ecliptic

Table 1. *The results displayed by* display_position *for various classes of objects. N signifies correction for nutation, A for aberration, G for geocentric parallax, and L for light time*

|            | True          | Apparent | Topocentric | Astrometric |
|------------|---------------|----------|-------------|-------------|
| Sun        | no corrections | N,A      | N,A,G       | –           |
| Moon       | –             | N        | N,G         | –           |
| Planets    | no corrections | L,N      | –           | –           |
| Elliptical | no corrections | –        | –           | L           |
| Parabolic  | no corrections | –        | –           | L           |

longitude, latitude, right ascension, and declination include corrections for geocentric parallax as well.

For the Moon, the displayed quantities are the *apparent* geocentric ecliptic longitude, latitude, right ascension, and declination which have been corrected only for nutation. The equatorial horizontal parallax, Earth–Moon distance, *apparent topocentric* right ascension, and declination have all been corrected both for nutation and geocentric parallax. Other quantities displayed include the geocentric angular diameter, the topocentric angular diameter, the phase, the position angle of the bright limb, the parallactic angle, and the vertical angle of the bright limb.

When displaying the results of a calculation made for a planet, the quantities are the *true* ecliptic longitude, latitude, right ascension, and declination which have been corrected for neither light time nor aberration. The Sun–planet radius vector, the Earth–planet distance, the *apparent* longitude, latitude, right ascension, and declination have all been corrected for light time and aberration. Other quantities displayed include the solar elongation, the angular diameter, the phase, the position angle of the bright limb, the parallactic angle, and the vertical angle of the bright limb.

If you specify an elliptical or parabolic calculation, you are asked to supply a name for the object, and then its orbital elements. You may give the perihelion position either as its argument (measured from the rising node of the orbit on the plane of the ecliptic – see Figure 19) or its longitude (measured from the vernal equinox), and the body's position as its mean anomaly (see Figure 20) or its longitude as convenient. For elliptical calculations, you also have the choice of specifying the period of the orbit or the mean daily motion (i.e. how many degrees it travels in its orbit in one day on average). You need also to specify the date of the equinox on which the elements have been based, and the date of the equinox for which you wish the position to be calculated. For example, the elements might be valid for the equinox of date 0.9,1,1950 (often abbreviated B1950), but you wish to calculate the position for the equinox of 1.5,1,2000 (J2000). The displayed quantities are the *true* heliocentric longitude and latitude with no corrections for light time, aberration or nutation, the Sun–object radius vector, the Earth–object distance, the *astrometric* ecliptic longitude, latitude, right ascension and declination (corrected for light time only), and the solar elongation of the object.

The simple two-line AstroScript program described above is direct and easy to understand, but you can easily write a more-general AstroScript program which calculates and displays the position of any orbiting object of your selection. See, for example, the program *orbits.txt*, which is on the disk which comes with the book. The program is as follows:

```
{Clear the screen, and put up a title}
display_title(*** Orbital Positions ***)

{Set the default argument for first run through}
if (s1=.)
  set_string(s1,Sun)
end_if
```

```
{For which object shall we make the calculation?}
ask_for_string(Calculate the position of which body,s1)

{Use the character variable c1 to flag an error}
set_character(c1,N)

{Check for allowed calculation types}
if (s1=sun)
   set_character(c1,Y)
end_if
if (s1=moon)
   set_character(c1,Y)
end_if
if (s1=mercury)
   set_character(c1,Y)
end_if
if (s1=venus)
   set_character(c1,Y)
end_if
if (s1=mars)
   set_character(c1,Y)
end_if
if (s1=jupiter)
   set_character(c1,Y)
end_if
if (s1=saturn)
   set_character(c1,Y)
end_if
if (s1=uranus)
   set_character(c1,Y)
end_if
if (s1=neptune)
   set_character(c1,Y)
end_if
if (s1=elliptical)
   set_character(c1,Y)
end_if
if (s1=parabolic)
   set_character(c1,Y)
end_if

{Calculate the position and display the result if allowed}
if (c1=Y)
   find_position(s1)
   display_position(s1)
end_if

{Raise an objection if the requested calculation is illegal}
if (c1!Y)
   set_normal_text(12)
```

```
    new_line
    write_string( +++ )
    write_string(s1)
    write_string( is not an allowed calculation type)
    new_line
    new_line
    set_normal_text(7)
end_if

{Another calculation?}
ask_for_repeat
```

When you run this script, you are asked for the calculation type and you must enter 'sun', or 'moon' etc. Note the use of *if . . . end_if* commands to check for a legal entry. Were you to enter 'fred' say, the value of the character variable c1 would not have been set to 'Y' by the time program execution reached the line *if (c1!Y)*, and an error message would then be printed. Let us try this program now to repeat our calculation for the position of Venus at $0^h$ on April 21st 1984, first using osculating elliptical elements, and then using the full perturbation calculation. We enter *ascript orbits.txt* at the DOS command prompt and press the *enter* or *return* key:

```
                    *** Orbital Positions ***
                    ------------------------

    Calculate the position of which body ........ ? elliptical

    Name of orbiting object ..................... ? Venus


    Eccentricity of the orbit (0-0.9999) ........ ? 0.006778
    Semi-major axis or mean distance (AU) ........ ? 0.723332
    Inclination of the orbit (degrees) .......... ? 3.394535
    Longitude of ascending node (degrees) ........ ? 76.589820

    Input perihelion argument or longitude (A/L) . ? L
    Longitude of the perihelion (degrees) ........ ? 131.430236

    Mean anomaly/longitude specified at G. date .. ? 0,1,1990
    Input mean anomaly or longitude at epoch (M/L) ? L
    Longitude at this epoch (degrees) ............ ? 88.455855

    The period/daily motion can be deduced from the size of the semi-major axis
    Deduce period/d. motion or input a value (D/V) ? V
    Input period or daily motion (P/D) ........... ? p
    Orbital period (years) ...................... ? 0.615211

    Elements specified at Greenwich date ......... ? 0,1,1990
    ..but make calculation for equinox of date ... ? 21,4,1984

    Calculate position at local calendar date...
    Calendar date (d,m,y; BC neg) ................ ? 21,4,1984
    Local civil time (h,m,s) ..................... ? 0,0,0
    Time zone (h ahead of UT; W negative) ........ ? 0
    Daylight saving (h ahead of zone t) .......... ? 0
```

```
                        ** The Position of Venus **
                        ---------------------------

     21-Apr-1984 in time zone: 0 with daylight saving: 0
     Local civil time: 0 0 0; UT: 0 0 0; ET/DT: 0 0 55

     Venus's true heliocentric longitude (d,m,s) ..    355 30  1.3630
     Venus's true heliocentric latitude (d,m,s) ...  -   3 21 10.5831
     Sun-Venus radius vector (AU) .................      0.7268346

     Venus's ecliptic longitude (d,m,s) ...........     16 14 44.5472
     Venus's ecliptic latitude  (d,m,s) ...........  -   1 28 33.0023
     Earth-Venus radius vector (AU) ...............      1.6506307
     Venus's astrometric right ascension (h,m,s) ..      1  2  8.8570
     Venus's astrometric declination (d,m,s) ......      5  1 36.8591

     Venus's solar elongation (d,m) ...............     14 54

     Again (Y or N) ............................... ? n
```

As before, the entries made via the keyboard are underlined. Compare these results with those obtained laboriously by hand above in the section *Low-precision orbit calculations*. The only difference between the two is that here we have made a correction for light time, and have solved Kepler's equation properly.

Now let us repeat the calculation using the full perturbation theory. Note that the offset of dynamical time from universal time is 55 seconds in this instance. Hence, in order to make the calculation for $0^h$ *dynamical* time on 21st April 1984, we must run the program for $23^h 59^m 05^s$ UT on 20th April 1984 as follows:

```
                        *** Orbital Positions ***
                        -------------------------

     Calculate the position of which body ......... ? Venus

     Calendar date (d,m,y; BC neg) ................ ? 20,4,1984
     Local civil time (h,m,s) ..................... ? 23,59,05
     Time zone (h ahead of UT; W negative) ........ ? 0
     Daylight saving (h ahead of zone t) .......... ? 0
     Geographic longitude (d,m,s) ................. ? 0 2 5.6700
     Geographic latitude  (d,m,s) ................. ? 52 10 12.0000
     Height above mean sea level (m) .............. ? 20.0

                        ** The Position of Venus **
                        ---------------------------

     20-Apr-1984 in time zone: 0 with daylight saving: 0
     Local civil time: 23 59 5; UT: 23 59 5; ET/DT: 24 0 0     {same as 0^h on 21st}

     Venus's true ecliptic longitude (d,m,s) ......     16 13 28.8417
     Venus's true ecliptic latitude  (d,m,s) ......  -   1 28 35.2855
     Sun-Venus radius vector (AU) .................      0.7268428
     Venus's true right ascension (h,m,s) .........      1  2  4.2379
     Venus's true declination (d,m,s) .............      5  1  5.7333
     Earth-Venus distance (AU) ....................      1.6502587
     Venus's apparent ecliptic longitude (d,m,s) ..     16 12 29.8066
```

```
Venus's apparent ecliptic latitude  (d,m,s) ..   -    1 28 35.6007
Venus's apparent right ascension (h,m,s) .....        1  2  0.5782
Venus's apparent declination (d,m,s) .........        5  0 43.7695

Venus's solar elongation (d,m,s) .............       14 55 51.0912
Venus's angular size (d,m,s) .................        0  0 10.1075
Venus's phase (full = 1.0) ...................        0.97
Position angle of bright limb (d,m; N=0) .....       61 55
Venus's parallactic angle (d,m) ..............   -    9 43
Vertical angle of bright limb (d,m; N=0) .....       71 38

Again (Y or N) ............................... ? n
```

The *Astronomical Ephemeris* reports the position of Venus as follows:

*Geocentric coordinates for $0^h$ dynamical time on 21st April 1984:*
*Apparent right ascension:* $1^h 02^m 00.59^s$
*Apparent declination:* $5° 00' 43.6"$
*True geocentric distance:* *1.6502861 au*

AstroScript is really quite accurate!

The command *display_position* is all very well, but often you do not want to display a whole page of information every time, but are interested only in the body's position specified in a particular coordinate system. AstroScript allows you to do this using an extension of the *convert_coordinates(⟨char1⟩, ⟨char2⟩)* command, together with the *set_coordinates_type(⟨string⟩)* command. The extension allows you to set the left-hand argument, *⟨char1⟩*, to one of 's', 'm', 'p', or 'o'. In each case the previously calculated right ascension and declination of the *S*un, *M*oon, a *p*lanet, or an *o*sculating position, respectively are converted into the coordinates specified by the right-hand argument ('h', 'r', 'a', 'e', or 'g' as described in the section *Converting coordinates using AstroScript*). By default, topocentric coordinates are used for the Sun and the Moon, apparent coordinates for the planets, and astrometric coordinates for the other bodies. However, you can specify otherwise by using the auxiliary command *set_coordinates_type(⟨string⟩)* in which the string argument can be one of 'true', 'apparent', or 'topocentric'. Table 2 shows which coordinates are converted in each case:

Table 2. *Coordinates converted by* convert_coordinates *for given arguments of* set_coordinates_type

|  | Sun | Moon | Planets | Osculating |
|---|---|---|---|---|
| 'true' | true | apparent | true | astrometric |
| 'apparent' | apparent | apparent | apparent | astrometric |
| 'topocentric' | topocentric | topocentric | apparent | astrometric |

For example, the following lines find and display the Moon's topocentric azimuth and altitude:

```
find_position(moon)
set_coordinates_type(topocentric)
convert_coordinates(m,a)
write_string(The Moon's topocentric horizon coordinates are:)
new_line
display_coordinates(a)
```

## Computing ephemerides

It is often very useful to produce a list of the positions of a given orbiting body at each of a regular sequence of instants. Such a list is known as an *ephemeris*, and AstroScript can produce and display one using the single command *compute_ephemeris(⟨string⟩)*. The argument ⟨*string*⟩ can be set to any one of 'sun', 'moon', 'mercury', 'venus', 'mars', 'jupiter', 'saturn', 'uranus', or 'neptune'. Thus, to see a list of the positions of the Moon, say at hourly intervals over the next day, you merely need to write

```
compute_ephemeris(moon)
```

in your AstroScript program. It is not possible to produce an ephemeris for osculating orbits using this command.

## Inverse position problems

The *find_position* calculations described above determine the positions of the given objects at a particular moment. For example, you supply the time and the date to *find_position(sun)*, and it returns with the Sun's position calculated for that instant. But often you wish to perform the calculation the other way around, finding the time and date at which the given object arrives at a particular position. For example, when does the Moon cross the meridian, or when is the ecliptic longitude of Jupiter exactly 180 degrees? AstroScript provides you with a particularly simple method of answering such questions with the *find_when(⟨string1⟩, ⟨string2⟩, ⟨string3⟩)* command. To calculate the moment when the Moon crosses the meridian, you would write

```
find_when(moon,hour angle,0)
```

or equivalently

```
find_when(moon,azimuth,180.0).
```

You set the first argument, ⟨*string1*⟩, to the name of the object, the second argument, ⟨*string2*⟩, to the coordinate type, and the third argument, ⟨*string3*⟩, to its target value.

The command returns with the date and time when the ⟨*string2*⟩ of ⟨*string1*⟩ reaches ⟨*string3*⟩. The first argument can be set to any of 'sun', 'moon', 'mercury', 'venus', 'mars', 'jupiter', 'saturn', 'uranus', or 'neptune' to make the calculation for that particular body, or to 'elliptical' or 'parabolic' to make an osculating orbit calculation. If you set this argument to anything else, the command assumes that you wish to make the calculation for the body whose right ascension and declination are currently held in the data bank, entered perhaps using the *ask_for_coordinates(r)* command. Thus, to find when a star's altitude reaches 11.23 degrees you could use the AstroScript program fragment

```
{Get the star's right ascension and declination}
ask_for_coordinates(r)

{Now find when the star reaches the desired position}
find_when(star,altitude,11.23)
```

The middle argument, ⟨*string2*⟩, can be set to 'hour angle', 'right ascension', 'declination', 'azimuth', 'altitude', 'ecliptic longitude', 'ecliptic latitude', 'galactic longitude', or 'galactic latitude'. The right-hand argument, ⟨*string3*⟩, must be a number, such as '12.5' or '237.2345' or '2.67e2' (i.e. 2.67 times ten to the power of two).

A variant of the inverse problem is to find when two bodies reach similar positions in the sky, or alternatively when they are exactly opposite each other. It is often important, for example, to find the instant when the Sun and the Moon share the same right ascension, or the same ecliptic longitude, i.e. when they are in *conjunction* with each other, or when their right ascensions differ by exactly 12 hours, or their ecliptic longitudes by 180.0 degrees, i.e. when they are in *opposition*. These questions can be answered in AstroScript using a slightly different form of the *find_when* command. This time, both the first and the third arguments, ⟨*string1*⟩ and ⟨*string3*⟩, are set to object names, and the middle argument, ⟨*string2*⟩, is set to one of 'conjunction in el', 'conjunction in ra', 'opposition in el', or 'opposition in ra' for conjunction in ecliptic longitude, conjunction in right ascension, opposition in ecliptic longitude, or opposition in right ascension, respectively. For example, to find the instant when the Moon and the Sun are in opposition in right ascension you would use *find_when (moon,opposition in ra,sun)*, or when the planets Mars and Jupiter share the same ecliptic longitude you would write *find_when(jupiter,conjuction in el,mars)*. As with the *find_position* command, the type of coordinates used in the search (true, apparent, or topocentric) can be specified with *set_coordinates_type( ⟨string ⟩)*.

The command *find_when* performs its calculations by making an iterative search for the answer. You must provide it with a starting point for this search, and it then returns with the answer closest to that moment. Give it another starting point to find the answer for another day (in the cases of hour angle, azimuth, and altitude) or for another cycle in orbit (say for right ascension etc.). If it cannot find the answer within a reasonable time, it aborts the calculation and returns with an error message. The maximum number of iterations it makes before giving up can be set by the *set_number_of_iterations(⟨string⟩)* command. The default value is 100, but you can set this to any other positive number within reason. Thus to set the maximum number of iterations to 20 you would write *set_number_of_iterations(20)*.

Another parameter controlling the search is the *tolerance*. This specifies how close the calculated position has to be to the target position to be acceptable. The default value is 1.0e−6 (one millionth), but you can set this to another value using *set_tolerance(⟨string⟩)*, e.g. *set_tolerance(0.01)* to set the tolerance to one hundredth.

The third parameter controlling the search is the maximum step size allowed between one calculation and the next. This can be adjusted using the *set_maximum_step* *(⟨string⟩)* command. By default the maximum step size has the value of 10, corresponding to a maximum change of plus or minus 10 days in the date for which the next test calculation is made. You should make this smaller if the search appears to miss intermediate solutions, or larger to make the search faster for slowly varying quantities such as the right ascension of Uranus. As a rule of thumb, set the maximum step size to about a quarter of the period of the quantity that you are testing. For example, to find successive passages of the Sun across the meridian (period of 1 day) use *set_maximum_step(0.25)*, whereas to find when the right ascension of Jupiter is next 23h (period of 11.86 years = 4332 days) use *set_maximum_step(1083)*.

The AstroScript program *inverse.txt* is on the disk which came with the book, and is a general inverse-problem solver. The program is:

```
{Clear the screen and display a title}
display_title(+++ Inverse Position Problems +++)

{Set default values for first pass}
if (s1=.)

{First ask for the controlling parameters}
    set_string(s1,10)
    ask_for_string(Maximum step size [days],s1)
    set_maximum_step(s1)

    set_string(s1,100)
    ask_for_string(Maximum number of steps in the search,s1)
    set_number_of_iterations(s1)

    set_string(s1,1e−6)
    ask_for_string(Tolerance,s1)
    set_tolerance(s1)

    new_line
    write_string(Allowed coordinate types are true, apparent, and topocentric)
    new_line
    set_string(s1,true)
    ask_for_string(Type of coordinates to be used,s1)
    set_coordinates_type(s1)
    new_line
```

```
{Now set the defaults for the find_when command}
  set_string(s1,Sun)
  set_string(s2,hour angle)
  set_string(s3, )
end_if

{Get the first argument}
write_string(Valid first arguments are:)
new_line
write_string(-------------------------)
new_line
new_line
write_string(Sun, Moon, Mercury, Venus, Mars, Jupiter, Saturn, Uranus, Neptune)
new_line
write_string(Elliptical, Parabolic, or name of body with current RA and Dec)
new_line
new_line
ask_for_string(First argument [choose one of the above],s1)

{Get the second argument}
new_line
write_string(Valid second arguments are:)
new_line
write_string(--------------------------)
new_line
new_line
write_string(Hour angle, Right ascension, Declination, Azimuth, Altitude,)
new_line
write_string(Ecliptic longitude, Ecliptic latitude, Galactic longitude,)
new_line
write_string(Galactic latitude, conjunction in el, conjunction in ra,)
new_line
write_string(opposition in el, or opposition in ra)
new_line
new_line
ask_for_string(Second argument [choose one of the above],s2)
new_line

{Discriminate between the two variants of the find_when command}
set_character(c1,A)
if (s2=conjunction in ra)
  set_character(c1,B)
end_if
if (s2=conjunction in el)
  set_character(c1,B)
end_if
if (s2=opposition in ra)
  set_character(c1,B)
end_if
if (s2=opposition in el)
  set_character(c1,B)
end_if
```

```
{Get the third argument according to find_when variant}
if (c1=A)
   ask_for_string(Target value,s3)
end_if
if (c1=B)
   ask_for_string(Second object name [Sun/Moon/Venus etc.],s3)
end_if

{Perform the calculation, and display the result}
new_line
find_when(s1,s2,s3)

{Another go?}
new_line
ask_for_repeat
```

As an example of using *inverse.txt*, we can find the moment of conjunction between Venus and the Sun in December 1639 corresponding to the transit of Venus observed by Horrocks:

```
                    +++ Inverse Position Problems +++
                    ----------------------------------

Maximum step size [days] ..................... ? 50

Maximum number of steps in the search ........ ? 100

Tolerance .................................... ? 1e-6


Allowed coordinate types are true, apparent, and topocentric
Type of coordinates to be used .............. ? topocentric


Valid first arguments are:
--------------------------

Sun, Moon, Mercury, Venus, Mars, Jupiter, Saturn, Uranus, Neptune
Elliptical, Parabolic, or name of body with current RA and Dec

First argument [choose one of the above] ..... ? venus


Valid second arguments are:
--------------------------

Hour angle, Right ascension, Declination, Azimuth, Altitude,
Ecliptic longitude, Ecliptic latitude, Galactic longitude,
Galactic latitude, conjunction in el, conjunction in ra,
opposition in el, or opposition in ra

Second argument [choose one of the above] .... ? conjunction in el


Second object name [Sun/Moon/Venus etc.] ..... ? sun
```

```
Please input a starting point:
Calendar date (d,m,y; BC neg) ............... ? 1,1,1640
Local civil time (h,m,s) .................... ? 0,0,0

Time zone (h ahead of UT; W negative) ........ ? 0
Daylight saving (h ahead of zone t) .......... ? 0
Geographic longitude (d,m,s) ................. ? 0 2 5.6700
Geographic latitude  (d,m,s) ................. ? 52 10 12.0000
Height above mean sea level (m) .............. ? 20.0

The calculation converges on the epoch when:
The local calendar date is ...............      Sunday  4-Dec-1639
The universal time is (h,m,s) .............      17 54  5.0750
The local civil time is (h,m,s) ...........      17 54  5.0750
Estimated ephemeris/dynamic time is (h,m,s) ..  17 55 18.9235

Again (Y or N) ............................... ? n
```

It is possible to use the *compute_sky* command (see section on *Sky graphics*) to display the passage of Venus across the face of the Sun at this time. Note, however, that conjunction occurred after sunset in England, so you should specify an earlier time, say $15^h$ $15^m$ UT. Remember to use the 'P' key to turn off the artificial planetary scaling.

The *find_when* command normally produces five lines of output itself, beginning with *The calculation converges on the epoch when:* and ending with a note about the ephemeris/dynamical time offset (see the above example). Sometimes it is useful to suppress this output as, for instance, when creating a data file with a given format using the *display_data* and *copy_to_file* commands. You can achieve this using *set_messages($\langle string \rangle$)* with $\langle string \rangle$ set to *OFF*. If later in the program you need to reinstate the output, use *set_messages(ON)*.

## Real-time programming

It is often necessary to write programs which display results continuously, updating values on the screen at regular intervals. Any AstroScript program can be run in this way, but the facility is especially useful for displaying the positions of orbiting objects whose equatorial coordinates are constantly changing, or for displaying the azimuth and altitude of any celestial object. The command *compute_sky* is an all-inclusive program in a single line which displays many objects at once (see later), but you may want to make up your own, more modest, program which displays the positions in real time of just one or two objects. You can do this using the commands *wait_time_step($\langle string \rangle$)* and *repeat*.

The *wait_time_step($\langle string \rangle$)* command suspends the execution of the program until the system clock has reached a time equal to or later than the last-used local civil time plus $\langle string \rangle$ seconds. The argument $\langle string \rangle$ must resolve to a number which is treated as positive whatever its sign, rounded to the nearest integer value, and interpreted as that number of seconds to be added to the last local civil time. Thus *wait_time_step(10)* adds 10 seconds, and *wait_time_step(3600.4)* adds exactly one hour (the extra 0.4 is ignored). The local civil time is set to the new value, and then execution continues with the next AstroScript command. If the waiting time spans local midnight, the date is changed too.

The *repeat* command simply causes the program to begin again from the top *after completing the final command*. It should, therefore, be the final command itself, though the interpreter will not object unless it is nested inside an *if . . . end_if* clause.

When writing real-time AstroScript programs, you should bear in mind that your computer may take some time to make a calculation. It is usually best therefore to let it complete all its calculations before clearing the screen ready to display the new values. For example, the following simple script displays the Moon's azimuth and altitude, updated every second:

```
display_title(The Moon's Position)
find_position(moon)
convert_coordinates(m,a)
display_coordinates(a)
wait_time_step(1)
repeat
```

It does not work very well, especially if your computer is a bit slow, because the screen remains blank while the computer is busy. The result is an irregularly flashing display. Much better results are obtained if you move the *display_title* command to the line immediately before the *display_coordinates* command. This ensures that the screen is blank for as short a period as possible:

```
find_position(moon)
convert_coordinates(m,a)
display_title(The Moon's Position)
display_coordinates(a)
wait_time_step(1)
repeat
```

Now the screen is blank for the minimum period, and the flashing effect much reduced.

Included on the disk with the book is a more-general real-time program example called *realtime.txt* as follows:

```
{Program to display an object's position in real time}

{Set the delimiter to '%' so we can use ',' in prompts}
set_delimiter(%)

{On first pass, get the name of the object etc.}
if (s1=.)
  set_string(s1%Sun)
  display_title(+++ Real-time Positions +++)
  write_string(Possible objects are:)
  new_line
  new_line
  write_string(Sun, Moon, name of planet, parabolic or elliptical)
  new_line
```

```
        new_line
        ask_for_string(Name of object[choose one of the above]%s1)
        set_string(s2%true)
        ask_for_string(Coordinates: true, apparent, or topocentric%s2)
        set_coordinates_type(s2)
        set_string(s3%5)
        ask_for_string(Time step between updates%s3)
        new_line
    end_if

    {Use the character variable c1 for both coordinate type and error}
    set_character(c1%N)

    {Check that the calculation is allowed}
    if (s1=sun)
        set_character(c1%s)
    end_if
    if (s1=moon)
        set_character(c1%m)
    end_if
    if (s1=mercury)
        set_character(c1%p)
    end_if
    if (s1=venus)
        set_character(c1%p)
    end_if
    if (s1=mars)
        set_character(c1%p)
    end_if
    if (s1=jupiter)
        set_character(c1%p)
    end_if
    if (s1=saturn)
        set_character(c1%p)
    end_if
    if (s1=uranus)
        set_character(c1%p)
    end_if
    if (s1=neptune)
        set_character(c1%p)
    end_if
    if (s1=elliptical)
        set_character(c1%o)
    end_if
    if (s1=parabolic)
        set_character(c1%o)
    end_if

    {Raise an objection if the requested calculation is illegal...}
    if (c1=N)
```

```
      set_normal_text(12)
      new_line
      write_string( +++ )
      write_string(s1)
      write_string( is not an allowed calculation type)
      new_line
      new_line
      set_normal_text(7)

{…wait for a response}
      pause

{…and reset s1 to its default condition}
      set_string(s1%.)

end_if

{If the calculation is allowed…}
if (c1!N)

{…carry out the calculation}
      find_position(s1)

{…convert to az/alt}
      convert_coordinates(c1%a)

{…clear the screen and put up a title}
      set_normal_text(15)
      display_title(s1)
      set_normal_text(7)

{…display the time and date}
      set_decimal_places(0)
      display_date(calendar)
      display_time(solar)
      new_line

{…display the result}
      write_string(Coordinate type: )
      write_string(s2)
      new_line
      set_normal_text(14)
      display_coordinates(a)
      set_normal_text(7)
      new_line

{…and wait for the clock}
      wait_time_step(s3)

end_if
```

{Finally repeat the whole lot with no questions asked}
repeat

Note the use of the character variable $c_1$ to indicate both an error (it remains set to 'N' unless the object name is legal), and to indicate the coordinates (s, m, p, or o) to be converted. Also note, no questions are asked when the calculation is repeated as $s_1$ is not equal to a point.

# Sky displays

Perhaps the best way of portraying the heavens is to use the computer screen in its graphical mode to paint a picture. You have to imagine that the screen represents the view through a telescope which you can point in any direction, and whose magnification you can vary at will, zooming in until the disk of Jupiter fills the field of view, or zooming out until you can see half the sky at once. The computer 'telescope' is especially flexible since you can take it anywhere in the world and see the sky at almost any instant in the future or the past, all without stepping outside the door. But in order to do this, we need a method of taking the real disposition of celestial objects on the inside *curved* surface of the celestial sphere, and placing them on the inside *flat* surface of the computer screen, in such a manner as to be as realistic as possible. The process is called a *mapping transformation*.

AstroScript incorporates one particular mapping transformation which it uses in the commands *compute_eclipses* and *compute_sky* (see below). Figure 22 shows a diagram of your computer screen when looking towards a point in the sky between north and east. The origin of the axes, O, represents the direction of alignment of the computer telescope, and Q' represents a star at relative position $(x, y)$ within the field of view. All the celestial objects in the hemisphere centred on O are mapped onto the screen within the dashed circle marked *limit of map* which may lie inside (as here) or outside the screen boundary depending on the magnification or scale. These are points which are in front of your eyes. In the diagram, the scale is quite small with the whole of the sky in front of you is represented on part of the computer screen. The cardinal points north and east actually lie within the dashed boundary in this case. As you zoom in, the circle becomes larger and larger, and you can see less and less of the sky at once. The point marked Z' represents your zenith. If O was on the horizon, then Z' would lie on the dashed circle. If you pointed the computer telescope at the zenith, then Z' would coincide with O, and the dashed circle would represent the horizon.

Figure 23 shows how the points on your computer screen relate the celestial sphere. You are to imagine that you are looking along the line through O to SC from below (your eye is shown at the bottom of the diagram). The computer screen is in the plane of the circle marked *plane of screen*; the circle itself corresponds to the dashed circle in Figure 22. SC is the point on the celestial sphere corresponding to the centre of the screen, and Q marks the position of a star. In this transformation, the image, Q', of the star at Q is found by dropping a perpendicular from Q onto the plane of the screen. Similarly, O represents the image of SC, and Z' represents the image of the zenith point, Z. The position of Q relative to SC is fixed by the angular distance, $\theta$, from SC

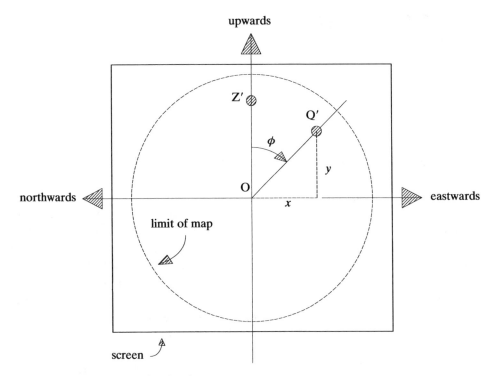

Figure 22. Mapping the sky onto your computer screen.

to Q, and by the angle, $\phi$, between the great circles through SC and Q, and through SC and Z. If $R$ is the radius of the celestial sphere, then

$$OQ' = R \sin \theta,$$

and hence

$$x = R \sin \theta \sin \phi,$$

$$y = R \sin \theta \cos \phi.$$

With these two equations, we can map any point such as Q on the celestial sphere onto the computer screen at the corresponding position $(x, y)$ provided that we know the values of $\theta$ and $\phi$. Note that we can alter the magnification merely by changing the value of $R$.

The point SC in Figure 23 can, of course, lie anywhere on the celestial sphere, and the general case is shown in Figure 24. This diagram is quite difficult to visualise, and you may find it helpful to see how the points in Figure 23 correspond with those in Figure 24 by turning back and forth between the two diagrams. Here, the horizon circle is shown with its north point, N. The direction of SC is then fixed by specifying its azimuth, $A_o$, and altitude, $a_o$. The corresponding coordinates for the star Q are $A$ and $a$. The formulae connecting $(\theta, \phi)$ with $(A_o, a_o)$ and $(A, a)$ are as follows:

$$\cos \theta = \sin a \sin a_o + \cos a \cos a_o \cos (A - A_o),$$

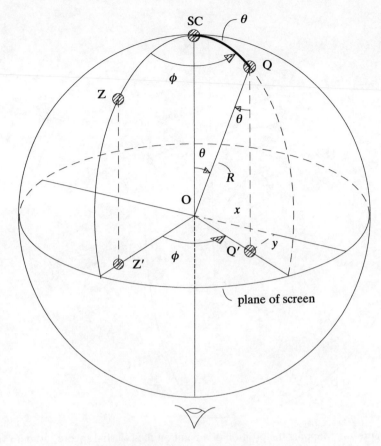

Figure 23. Your computer screen and the celestial sphere.

and

$$\tan \phi = \frac{\sin (A - A_o)}{\cos a_0 \tan a - \sin a_o \cos (A - A_o)}.$$

These are the formulae used by AstroScript to represent the sky on the computer screen.

## *Exploring the sky with AstroScript*

The single most powerful command in AstroScript is *compute_sky*. This incorporates nearly every other command to provide you with an 'astronomical desk' from which you can explore the sky. The AstroScript program is especially simple, and

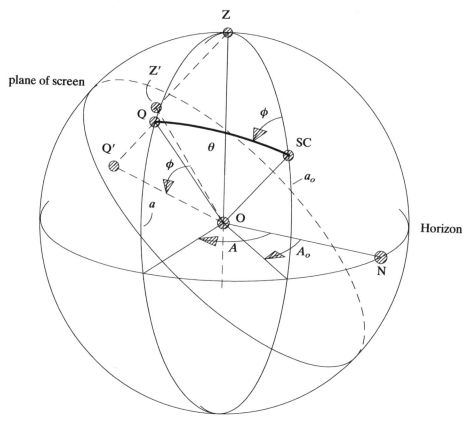

Figure 24. Mapping the position of a point on the celestial sphere, $Q$, onto your computer screen at $Q'$.

is called *ast_desk.txt* on the disk which comes with the book:

```
{Clear the screen, and display a title}
display_title(+++ Astro Desk +++)

{Invoke the astro-desk command}
compute_sky

{Another go?}
ask_for_repeat
```

When you run this program (by entering *ascript ast_desk.txt* and pressing the *enter* or *return* key), you are first asked if you wish to use a colour or monochrome monitor. Enter *C* for colour unless you are using a black and white screen without any grey scales. You are then given the choice of seeing the star list. The command *compute_sky* incorporates the facility of displaying stars and other objects of your choice, and you can edit the default list of objects by replying *Y* to the question 'See the star list (Y or N) ?'. We shall return to this later, but for now reply *N*. You are immediately given a list of keys which remain active during the running of the program. as follows:

```
Active keys:
------------

Press...
RETURN        to enter a new geographical position
S             to get time/date from system clock (default)
U             to enter your own time and date (user time)
P             to display the other page(s)
D             to display a view of the sky as seen from the ground
Q or ESCAPE to exit

Page 1 shows the positions of Solar-System objects
Page 2 shows their circumstances of rising and setting
Page 3 shows the positions of other stars
Page 4 shows their circumstances of rising and setting
```

You can change the geographical position of the observer by pressing the *return* or *enter* key. Pressing *S* (for 'system time', the default setting) causes the program to run continuously in real time, getting the local civil time and date from the system clock and updating the values on the screen every second (if your computer is fast enough). Pressing *U* (for '*u*ser time') sets the local civil time and date to a value which you provide from the keyboard, and the program remains fixed at that time until you press *S* or *U* again. There are four continuously updated pages of tables of text information available to you. Page 1 is displayed by default, but you can see the others by pressing *P* to go forward to the next *p*age. Pressing *P* when on page 4 takes you back to page 1 again. You can also obtain a graphical *d*isplay of the sky by pressing *D* (see below). Press *Q* or the *esc* key to exit from *compute_sky*.

Now press any key to start the program at page 1. You are asked for your geographical longitude, latitude, time zone, and daylight saving correction. Just press return for the default values as displayed, or enter other values if desired. You can repeat this process at any time whilst the program is running by pressing the *enter* or *return* key as mentioned above. Having entered the daylight saving correction, the program begins running in the mode shown in the top left-hand box on the screen. By default, this is the real-time mode in which the system clock provides the local civil time and date. The four other boxes on the top row display the corresponding times, and the two boxes side-by-side on the right-hand side of the screen below the top row show the Julian and calendar dates. Immediately below them is another box showing the page number (1 to 4) and the title of the page. The large box occupying the lower half of the screen displays the astronomical information. Pages 1 and 2 give information for the Sun, Moon, Mercury, Venus, Mars, Jupiter, Saturn, Uranus, and Neptune. On page 1 you have the right ascension and declination, the azimuth and altitude, the distance from Earth, and the phase. An example of what you might see is overleaf.

Press *P* for page 2 and see the circumstances of rising and setting for these objects.

Pressing *P* again takes you on to page 3. Here you see the right ascension, declination, azimuth, and altitude of the stars in the star list, displayed 9 at a time. These are precessed coordinates, correct for the current date and time. Press + to move down the list for the next 9, and − to move back up the list. The number in the list of the first object displayed is shown in the title box. The names of the objects in the list (if you have provided them) are displayed in the left-hand column. The file

```
System      G Sidereal Time  L Sidereal Time  Universal Time   Local Civil Time
 time            9 17 12.8        9 17 21.2       20  6 50.0        20  6 50.0


Geographic longitude (d,m,s)  0 2 5.7          Julian Date     Local Civil Date
Geographic latitude (d,m,s) ? 52 10 12.0                           Sunday
Time zone (h ahead of UT; W negative) 0       J  2449818.18      9-Apr-1995
Daylight saving (h ahead of zone t) ? 0


                                          SOLAR SYSTEM OBJECTS: page    1


OBJECT    RA (H,M,S)    DEC (D,M,S)    AZ (D,M,S)     ALT (D,M,S) DIST (AU) PHASE
Sun        1 12  0.9     7 37 48.7    299 59 17.7   - 12 11 44.7  1.001655
Moon       8 34 29.2    12 39 15.3    196 14 28.1     49 33 12.6 393358 km  0.66
Mercury    0 55 26.0     4 25 39.7    301 51  8.4   - 17  5 39.7  1.340561  0.98
Venus     23  7  6.3  -  6 56  4.2    323 39 40.5   - 39 28 21.1  1.284978  0.81
Mars       9 11 53.5    19  6 58.1    182 21 28.0     56 55 44.3  0.968192  0.92
Jupiter   16 56 27.2  - 21 48  6.4     84 36 27.1   - 32  8 29.4  4.726777  0.99
Saturn    23 22 58.7  -  5 57 39.7    319 29 33.7   - 37  3 32.6 10.505056  1.00
Uranus    20 10  2.9  - 20 37 10.7     28 51 11.9   - 55 51 16.8 19.869629  1.00
Neptune   19 49 16.6  - 20 30 20.3     36 42  7.3   - 54  1 22.3 30.261755  1.00
```

*stardata.asc* which is provided on the disk with the book contains the details of about 1600 bright stars. You can add to this list, or modify it, either using AstroScript or with a text editor (see below). Press *P* again to display the circumstances of rising and setting for these objects (9 at a time).

You can return to page 1 by pressing *P* another time. Now try entering a different time and date. Press *U* and you will see the position box change colour from cyan to magenta, and extend downwards to provide space for you to enter the local civil time and calendar date. Try the values 0,0,0 for the time (midnight) and 1,1,2000 for the date. The mode box (top left) now displays 'User time' and the other boxes remain static, giving the information for the entered instant only. You can range through the pages as before using the *P* key. You can also change your position on the Earth by pressing *enter* or *return*. If you now enter 0,0,0 for the geographical longitude, 89,59,59 for the latitude, and zeros for the time zone and daylight saving correction, you will see the positions of the objects as seen from the north pole at the turn of the century. Note that it is always good practice to avoid using 90,0,0 or −90,0,0 for a latitude (geographical, ecliptic, or galactic) since the equations become ill-conditioned at those points and, although AstroScript may not produce an error, the results should not be trusted. Press *S* to return to the real-time mode.

## *Sky graphics*

Another function provided by the AstroScript command *compute_sky* is a graphical display of the sky, i.e. the 'computer telescope' mode described at the beginning of this chapter. You can invoke this by pressing *D* at any time when the program is running. The screen clears and, if your machine does not have a coprocessor, you are given a warning that the response might be too slow to be useful. I have not attempted to degrade the inherent accuracy of AstroScript to speed up graphics displays, so you might need to be patient. A 486DX50 machine takes no more than a couple of seconds to draw the screen; a 486SX25 without a coprocessor, on the other hand, takes about 33 seconds. The software will work on *any* 8086 class of machine which has a VGA screen, but might be very slow without the corresponding 8087 coprocessor. If you reply *Y* to the question 'Are you sure you wish to proceed . . . ?', you are given a list of the keys which remain active during the graphics display mode as follows:

```
Active keys during display mode:
--------------------------------

Press...

H    for help

arrow keys, followed by ENTER, to change view direction in small steps
Home/end/Page Up/Page Down, followed by ENTER, to change in large steps
I/O to zoom in/out; 1 to 9 for preset zoom settings [default 1]
G    to enter a new geographical position
S    to get time/date from system clock [default]
U    to enter your own time and date (user time)
M    to toggle display of solar-system objects [default ON]
W    to toggle display of stars [default ON]
E    to toggle display of equator, ecliptic and galactic plane [default ON]
L    to toggle the star labels on/off [default ON]
C    to toggle the constellation lines on/off [default ON]
P    to toggle true scaling of planet disks on/off [default OFF]
A    to toggle atmospheric correction on/off [default OFF]
Q    to exit
```

Some of the keys act more or less as they did before, such as 'S', 'U', and 'Q'. However, 'G' now replaces *enter* or *return* for entering a new geographical position, and you have additional keys for changing the magnification, changing the viewing direction, switching the display of various features on and off, and invoking corrections for atmospheric refraction and planet scaling. You can bring up a help window by pressing the 'H' key (see below).

You can zoom in or out in steps by pressing *I* or *O* repeatedly. Alternatively, you can go straight to one of nine preset settings by pressing a digit key, i.e. *1* for setting number one, *5* for setting number five, or *9* for setting number nine, and so on. Setting 1 corresponds to a zoom factor of 1, and setting 9 to a zoom factor of 50. The other settings give zoom factors in between. The viewing direction can be altered using the arrow keys for small steps, or 'page up', 'page down', 'home', and 'end' for large steps, followed by 'enter', as explained below.

Press any key to start the sky display.

Having calculated the various elements of the view to be displayed, such as the positions of the planets, curve of the ecliptic plane etc., the computer draws a view of the sky looking due south along the horizon with a magnification (or zoom factor) of 1.

The position of the 'screen centre', SC in Figures 23 and 24, is shown by the cyan-coloured cursor. Imagine that you are within a huge dome (like a planetarium) whose inside surface is marked with lines of constant altitude (going from side to side) and curves of constant azimuth (going from top to bottom). The horizon is the green horizontal line near the bottom of the screen, representing zero altitude. Parallel to it are dotted grey lines representing altitudes of 10°, 20°, 30° etc. The cursor is in the centre of the horizon line, and its azimuth and altitude are displayed in the space just below it, along with the zoom factor and the local civil time. Below these is another space showing the active keys. Curves of constant azimuth are drawn as dotted grey lines at 10 degree intervals stretching upwards from the horizon and meeting at the zenith (which is above the top of the screen at the moment). The most easterly curve at the left-hand side of the screen represents azimuth 90°, and the most westerly curve at the right-hand side of the screen represents 270°. Thus almost the whole hemisphere in front of your eyes is mapped onto the computer screen.

If your computer has sufficient unused memory, you can bring up a help window showing the actions of the various active keys by pressing the *H* key. If there is not enough memory, you hear a beep instead and the request is ignored.

Zoom in a bit by pressing the *I* key. Now the magnification is 1.3, and you can see less of the sky at once (though in greater detail). Press *I* again to zoom in further, or *O* to zoom back out again. Notice that a cyan-coloured asterisk appears in the top right-hand corner of the screen whenever the computer is busy making calculations. You can tell that it has finished when the asterisk disappears, but you need not wait to move to another zoom setting or to change the viewing direction, as pressing any key interrupts the processing and gives you immediate control. Press *1* to go back to the original zoom setting.

The arrow keys allow you to change the viewing direction in small steps, or 'page up', 'page down', 'home', and 'end' to change in large steps. Press the up arrow key four times. You will see that the cursor has moved up the screen by 4 degrees, and has changed colour from cyan to green, leaving behind it a blue image of itself at its old position. Now press *enter* or *return*, and the screen is redrawn centred on the new viewing direction. Don't be puzzled by the fact that the horizon line appears to rise up at the edges of the screen and curve back on itself, nor by the additional azimuth and altitude curves at the edges. The program is actually displaying some of the sky which lies behind you as well as that which is in front of you, and so you can imagine these wings at the edges are curving round *behind* you as you stand inside the planetarium. They are only apparent at zoom factor 1. Now press *page up* twice. This moves the cursor up by 60°, and when you press *enter* or *return* the screen is redrawn at the new position. (**Note that you must always press *enter* or *return* after moving the cursor to reactivate the screen.**) The zenith is near the top of the screen. Press *Home* to move in azimuth steps of −90 degrees, or *End* to move in steps of +90. Press either key twice, followed by *enter* or *return* to redraw the screen centred on azimuth 0° and altitude +64°.

There are three other curves displayed besides those of constant azimuth and altitude. The brown dashed curve represents the trace of the celestial equator, the red dashed curve represents the trace of the plane of the ecliptic, and the magenta dashed curve represents the plane of the Galaxy (the Milky Way). Objects displayed include the Sun (solid yellow disk), the Moon (white), Mercury (green), Venus (white), Mars (red), Jupiter (cyan), Saturn (brown), Uranus (magenta), and Neptune (light grey). The Moon and the planets are all shown with their correct phases, but you will need to point your telescope at them and zoom in to see them properly. Press *U* and set the local civil time to 16,0,0 and the date to 9,10,1994 (9th October), and *G* to set the time zone to 0 and daylight saving to 1. With the zoom factor set to 1 (press *1*), move the cursor until it is over Venus (azimuth about 199° and altitude about 12°). Press *enter* or *return* to redraw the screen, and press *9* to zoom in to maximum magnification. You should then see the crescent of Venus filling the screen. I have had to cheat a bit with the scaling for the planets since they are so small that a very large magnification is needed to see them properly. The distances between objects, and their sizes, vary linearly as the zoom factor, except for the planets whose radii vary as the *square* of the zoom factor. Thus, when you view Venus with the zoom factor set to 50, it appears on the screen magnified 2500 times, but with its centre at the correct position. If you wish, you can turn off (and on again) this artificial scaling by pressing the *P* key.

Other objects displayed include all the objects in the star list and the Earth's shadow (unfilled circle). The stars are displayed using filled white and grey disks of various sizes to simulate their different magnitudes, and other objects such as star clusters or galaxies can be displayed as open circles by specifying a magnitude of −100 (see below). All the objects in the star list can also be joined together with straight lines to represent the constellations. The Earth's penumbra and umbra are always drawn after the Moon so that lunar eclipses are correctly displayed. Try, for example, a user time (press *U*) of 5,10,0 (time zone 0 and daylight saving 1) on date 13,4,1968. Move the viewing centre to azimuth 242° 30′ and altitude 6° 57′ (using geographical longitude 0,2,5.67 and latitude 52,10,12 – the default settings on the disk). Now press *7* to set the zoom factor to 20. You should see an eclipse of the Moon in progress! The unfilled largest circle represents the Earth's penumbra, and the black inner region the Earth's umbra. This is the moment just 13 minutes before the start of the total phase. Change the time to 5,23,0 to see for yourself (but remember to repoint your telescope since the Moon moves a long way in 13 minutes). Try also viewing the millennium solar eclipse at about 11,15,0 on the 11th August 1999. You will need to point your telescope towards azimuth 139° and altitude 47° (use the default settings for geographical longitude and latitude), and set the zoom factor to 35 (press *8*). See later in this chapter for a discussion of eclipses.

## *The star list*

The AstroScript command *compute_sky* includes the facility for displaying the positions of up to 2000 stars and other objects of your choice. These are kept in a text file called *stardata.asc*, which can be edited using any text editor like DOS *edit* or Windows *Notepad*. AstroScript itself also incorporates a limited list editing facility. The first 11 lines in the default *stardata.asc* file supplied on the disk are:

```
1.50-Jan-2000
    1 Alp UMi,       2 31 48.7000,    89 15 51.0000,    2.0,      2,       0
   23 Del UMi,      17 32 12.9000,    86 35 11.0000,    4.4,      7,       0
   HD 005848,        1  8 44.7000,    86 15 25.0000,    4.3,      0,       0
   HD 217382,       22 54 24.8000,    84 20 46.0000,    4.7,      0,       0
   HD 216446,       22 47 29.0000,    83  9 14.0000,    4.7,      0,       0
   VZ Cam,           7 31  4.4000,    82 24 41.0000,    5.0,      0,       0
   22 Eps UMi,      16 45 58.1000,    82  2 14.0000,    4.2,      9,       0
   HD 081817,        9 37  5.2000,    81 19 35.0000,    4.3,      0,       0
   16 Zet UMi,      15 44  3.5000,    77 47 40.0000,    4.3,     20,       0
    1 Kap Cep,      20  8 53.3000,    77 42 41.0000,    4.4,      0,       0
```

The top line, 1.50-Jan-2000, sets the epoch for which the star data is valid. (Note that all the positions given in the file **must** be for the same epoch.) There can then be any number of lines up to 2000, one line for each object. The six fields in each line are delimited by commas, and represent the name of the object (first field: up to 10 characters), the right ascension (second field: hh mm ss), the declination (third field: dd mm ss), the visual magnitude (fourth field), join-to number 1 (fifth field) and join-to number 2 (sixth field). These last two fields, the 'join-to' numbers, are the object numbers in the table to which straight lines are to be drawn to represent a constellation. No lines are drawn for join-to numbers of zero. Thus, in the table above, the star 23 Del UMi is joined by a straight line to the 7th object in the list (on line 8) which is the star 22 Eps UMi, and this in turn is joined to star number 9, 16 Zet UMi. When you display the sky, you can see that these lines are part of the outline of constellation Ursa Minor. AstroScript corrects the positions of the objects for precession before displaying them, so that the picture drawn is appropriate to the date set. It does not, however, take account of proper motions.

As mentioned above, you can edit the star list as you please to remove or to add objects of your own choice. I very strongly advise that you first make a backup copy of the file *stardata.asc* before editing as then you can begin again from the beginning if you get in a muddle. If you use an editor such as DOS *edit*, you will notice that the number of each object in the list is one less than the line number displayed by the editor. This is because the editor counts the top line, the date of the epoch, as line number 1 and so the list of objects begins on line 2. You might therefore find it easier to use the AstroScript editing facility which you invoke using the line

```
edit_star_data
```

in an AstroScript program. (It is also included within the *compute_sky* command.) You will first be asked whether you wish to view the star list. Answer *y* and you will then be asked where in the list you wish to begin as follows:

```
Reading the star-file...
See the star-list (Y or N) ................... ? y
Start at which object number ................ ? 23
```

This produces a display of the objects in the list beginning with the page (12 lines each) containing the object number you specified:

```
              *** STAR LIST (epoch  1.50-Jan-2000) ***
              ------------------------------------------

(Page number    2)

            Name          Right Ascension   Declination  Magnitude Join-to numbers

        13: 15 The UMi:   15 31 24.9000    77 20 58.0000     5.0       0    0
        14: HD 049878 :    7  0  4.0000    76 58 39.0000     4.6       0    0
        15: 21 Eta UMi:   16 17 30.3000    75 45 19.0000     4.9       9    0
        16: HD 091190 :   10 35  5.5000    75 42 47.0000     4.8       0    0
        17:  5 UMi    :   14 27 31.5000    75 41 46.0000     4.3       0    0
        18: 33 Pi Cep :   23  7 53.9000    75 23 15.0000     4.4       0    0
        19: HD 019275 :    3 11 56.3000    74 23 37.0000     4.9       0    0
        20:  7 Bet UMi:   14 50 42.3000    74  9 20.0000     2.1      26    0
        21: 60 Tau Dra:   19 15 33.0000    73 21 20.0000     4.4       0    0
        22: 44 Chi Dra:   18 21  3.4000    72 43 58.0000     3.6       0    0
        23: 50 Cas    :    2  3 26.1000    72 25 17.0000     4.0       0    0
        24: 24 Cep    :   22  9 48.4000    72 20 28.0000     4.8       0    0

    Press C then <Enter> for next page, or Q to quit q
```

Press *C* then *enter* to see each subsequent page, or *Q* then *enter* to move on to the editing:

```
    Update the star list (Y or N) ................ ? y
    How many stars in the list (0 to 2000) ....... ? 1605
    Begin updating at which star (1-1605) ........ ? 1605
    Stop updating at which star (1605-1605) ...... ? 1605

    Please give details:

    Star number 1605: name (up to 10 characters) . ? New-obj 1
    Right-ascension (h,m,s) ...................... ? 1 2 3.0000
    Declination     (d,m,s) ...................... ? 23 34 45.0000
    Stellar magnitude ............................ ? -100.0
    Join to star numbers (0 for no join) ......... ? 0, 0

    These coordinates are for epoch ............. ? 1.5000/ 1/ 2000
    Display the values again (Y or N) ........... ? y
```

If you elect to update the star list (as here), you are asked to specify the length of the list. The default value is the current length; increase this by one to add an extra object. Always add new objects at the end of the list since otherwise you will throw out all the join-to references, and your constellations will look ridiculous! You can delete objects from the end of the list by specifying the list length to be shorter than its current value. For example, to remove object number 1605 in a list of length 1605, just specify a list length of 1604. If you specify the magnitude to be −100 (as here) the object will be displayed as an unfilled circle, useful if you wish to differentiate it from a star. Note that the epoch of the coordinates applies to ALL the objects in the list.

## The phases of the Moon

The relative positions of the Moon and the Sun as viewed from the surface of the Earth are always changing throughout the course of a month. The hemisphere of the Moon facing the Sun is brightly illuminated, but we can only see the hemisphere which faces the Earth. Unless the Moon is in opposition to the Sun, the time of *full moon*, our hemisphere is not uniformly illuminated but overlaps both the bright and dark hemispheres. Then we see only part of the disk, and the Moon takes on its characteristic crescent shape. The area of the bright part of the Moon, expressed as a fraction of the whole disk, is known as the *illuminated fraction* or the *phase*. However, the latter term is also used to describe the particular instances when the differences between the apparent longitudes of the Moon and the Sun are exactly 0°, 90°, 180°, or 270°, i.e. the times of *new moon*, *first quarter*, *full moon*, and *last quarter*, respectively, these being known collectively as the 'phases of the Moon'.

You can calculate the phase, $k$, of the Moon for any given instant by first finding the true longitude of the Sun, $\lambda_s$, the true longitude of the Moon, $\lambda_m$, and the true latitude of the Moon, $\beta_m$, and then using the formulae

$$a = R_s \sin \{\cos^{-1}[\cos \beta_m \cos (\lambda_m - \lambda_s)]\},$$

$$b = R_m - R_s \cos \beta_m \cos (\lambda_m - \lambda_s),$$

$$k = \frac{1 + \cos [\tan^{-1}(a/b)]}{2},$$

where $R_s$ and $R_m$ are the distances from Earth of the Sun and the Moon, respectively (measured in the same units of course). This is how the phase is calculated in the *find_position* command in AstroScript. The times of new moon, first quarter, full moon, and last quarter may also be found and displayed using the single command *compute_moon_phases*. The procedure used is based upon the method given by Meeus in *Astronomical Algorithms* (see *Bibliography*), and it calculates the instants of the true phase to within a few seconds of time.

## Eclipses

Every month at the moment of new moon, the directions of the Moon and the Sun lie quite close to one another. Indeed, if the Moon's apparent orbit about the Earth lay in the plane of the ecliptic, then those directions would coincide exactly, the Moon would be seen to pass in front of the Sun, and a *solar eclipse* would be observed. As it is, the Moon's orbit is inclined to the ecliptic at an angle of about 5°, so that the directions of the Moon and the Sun rarely coincide closely enough for the Moon's disk to overlap that of the Sun. Very occasionally, however, the moment of new moon occurs at about the same time as the Moon is passing through one of its nodes, and then a solar eclipse can occur. The rule is that the angle between the line of nodes and the position of the Sun at the moment of conjunction in ecliptic longitude must be less than 18° 31′ for there to be the possibility of a solar eclipse. If that angle is less than 15° 31′, a solar eclipse will definitely occur.

In a similar fashion, the Moon can sometimes pass through the Earth's shadow at the time of full moon. The shadow always lies in the plane of the ecliptic exactly opposite the Sun, and if the Moon passes through one of its nodes near to full moon, a *lunar eclipse* can occur. We then see a 'bite' missing out of the lunar disk which gradually increases in size. If the eclipse is total, the bite eventually engulfs the whole disk and the Moon dims to a faint reddish hue. If the angle between the line of nodes and the Moon at the moment of opposition in ecliptic longitude is less than 12° 15′, a total lunar eclipse can occur, other circumstances being favourable. If the angle is less than 9° 30′, a total lunar eclipse must happen.

The AstroScript command *compute_moon_phases* applies these rules and declares the definite, likely, or non-occurrence of both solar and lunar eclipses. Try, for example, the date 21st October, 1994. You will see that both a lunar and a solar eclipse occur during the lunation in progress on that date. (A new lunation begins at every new moon.) Whether you can observe either eclipse depends, however, on your position on the Earth.

A lunar eclipse begins with the *penumbral phase* when the Moon enters the penumbra (or outer part) of the Earth's shadow, and the Moon's disk becomes a little fainter. You probably would not notice this unless you were looking for it. In Figure 25(a), the Moon at position *A* lies entirely within the penumbra. The Moon may then go on to enter the umbra (inner part) of the Earth's shadow. This marks the start of the *umbral* or *partial phase*. Here, the direct paths of light rays from the Sun to the Moon are entirely blocked, and that part of the Moon's disk lying within the shadow looks very dark. The face of the Moon does not go completely black, however, as some light still reaches it by refraction through the Earth's atmosphere. Imagine that you were standing on the Moon looking towards the Earth just as the lunar eclipse began. You would see the Sun 'setting' behind the (larger) disk of the Earth. If there were a lot of dust in the atmosphere, the sunset might cast a deep red glow onto the Moon, and an observer on the Earth would see a dim coppery lunar face. If the circumstances of the eclipse are favourable, the Moon may pass entirely within the umbra, as at B in Figure 25(a). This is the *total phase* of the eclipse.

A solar eclipse begins with the partial phase when the Earth enters the penumbra of the Moon's shadow. The Moon's dark face is turned towards us (it is the time of new moon) and so you cannot see the Moon in the day-time sky, especially when it is so near to the Sun. Instead you see a 'bite' missing out of the Sun's disk which increases as the eclipse progresses. If you are favourably situated, you will see the Moon eventually cover the whole of the Sun's disk. The eclipse is then in its total phase. The sky goes dark, the birds stop singing, and spectators are hushed as they observe the heavens' most awe-inspiring phenomenon. Since the Moon is so much smaller than the Earth, its umbra extends a much shorter distance into space, in fact only just far enough to reach the Earth's surface when the conditions are right (Figure 25(b)). The tip of the umbra

(a)                                    (b)                                    (c)

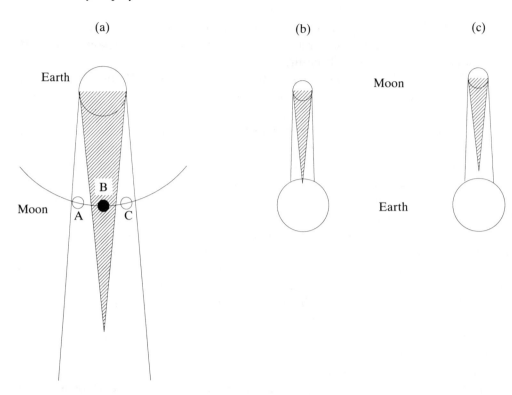

Figure 25. Eclipses: (a) Lunar eclipse. (b) Total solar eclipse. (c) Annular solar
eclipse.

casts a small shadow on the face of the Earth which moves across it as the relative
positions of Sun, Moon, and Earth change. Never is it sufficiently large to engulf the
whole Earth. Consequently, any total eclipse can only be seen along a narrow strip of
the Earth's surface. Sometimes, however, the umbra does not reach the Earth at all, as
in Figure 25(c). In this case, an annular eclipse can occur with the Moon not quite
obscuring the whole of the Sun's disk at maximum eclipse, but leaving a ring of light
around its edge.

It is one of the strange coincidences of nature that the angular size of the Moon, as
seen from the surface of the Earth, is just about the same as that of the Sun. Were it
much smaller, we would never observe total solar eclipses at all, but at best only broad
annular eclipses. Were it much larger, then the Moon's umbra would cover a greater
area of the Earth's surface, solar eclipses would be much more commonly observed at
any given place, and annular eclipses would never be seen at all. As it is, there is a fine
balance and we see both total and annular eclipses. It will not always be so, however,
as the Moon is very gradually moving away from the Earth. The action of the Moon's
gravity on the Earth causes the tides in the sea. We can imagine a great bulge of water
sloshing round the Earth after the Moon. But the Earth rotates once in twenty-four

hours, whilst the Moon orbits the Earth once in 28 days. The land masses are therefore constantly being forced through the tidal bulges, causing a drag which gradually slows down the Earth. The angular momentum of the Earth is, by this process, being slowly transferred to the Moon, with the result that the Moon's orbital radius is increasing. Far into the future, then, our descendants will no longer observe total solar eclipses but only annular ones.

## The 'rules' of eclipses

Here is a summary of the most important rules which appear to govern the occurrence of eclipses:

(1) A lunar eclipse can only occur at full moon, and a solar eclipse at new moon. There is not an eclipse every month because the Moon's orbit is inclined at an angle to the plane of the ecliptic.
(2) At least two solar eclipses occur every year, and never more than five. There is a maximum of three lunar eclipses in a year. A solar eclipse is therefore more common than a lunar eclipse, but can only be seen along a narrow strip of the Earth's surface whereas a lunar eclipse is visible over the entire hemisphere for which the Moon is above the horizon. Hence solar eclipses are quite rare at a given location.
(3) Eclipses tend to go in pairs or threes: solar–lunar–solar. A lunar eclipse is always preceded or followed by a solar eclipse (with two weeks in between).
(4) The pattern of eclipses tends to recur in cycles of 18 years 11 days and 8 hours, the so-called *Saros cycle*. The pattern does not repeat exactly.
(5) At the moment of greatest eclipse, the Sun and the Moon are either in opposition or conjunction in ecliptic longitude. If the angle between the line of nodes and the Sun or Moon is greater than 12° 15′ a total lunar eclipse is not possible, whilst if it is less than 9° 30′ a lunar eclipse *must* occur. If the angle is more than 18° 31′ a solar eclipse cannot happen, whilst if it is less than 15° 31′ a solar eclipse *must* occur.
(6) In a lunar eclipse, the total phase can last for a maximum time of 1 hour 40 minutes, and the umbral phases, partial–total–partial, for a maximum time of 3 hours 40 minutes. The maximum time of a total solar eclipse (at the equator) is 7 minutes 40 seconds, and an annular eclipse may not last for more than 12 minutes and 24 seconds.

## Calculating eclipses

The geometry of an eclipse is illustrated in Figure 26. The line AIHGF represents the plane of the ecliptic, and the line ABCDE represents the path of the Moon. The circle of radius $R_s$ centred on G represents the Sun in the case of a solar

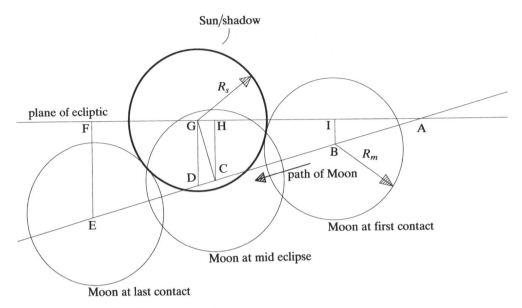

Figure 26. The geometry of an eclipse.

eclipse, or the Earth's shadow in the case of a lunar eclipse, either umbra or penumbra. The Moon is represented by the circles centred on B, C, and E. When at B, the Moon's disk just touches the Sun/shadow and the eclipse begins. Point C is the Moon's position of maximum eclipse, and E the position at last contact. Point D shows the position of the Moon at conjunction in ecliptic longitude (solar eclipse) or opposition in ecliptic longitude (lunar eclipse). We are to imagine that the position of G is fixed in the sky and that only the Moon is moving, though in reality both are moving at once. We can do this by subtracting the actual motion of point G from that of the Moon.

The eclipse calculation begins by finding the time (dynamical time) at which the Moon's latitude reaches zero (point A). Let this moment be called $DT_0$. We can imagine that the Moon advances steadily along its path during the eclipse, so we can convert distances on the diagram into time differences by multiplying by a constant scale factor. We can therefore find the time of first contact by adding the scaled distance AI to $DT_0$, and the time of mid eclipse by adding the scaled distance AH to $DT_0$, and so on. We calculate the distance AG by finding the position of the Sun/shadow at $DT_0$, and the distance GD by multiplying the rate of change of Moon's latitude by the time taken for the Moon to reach point D. Let AG be equal to $z$, let GD be equal to $s$, and let AH be equal to $\eta$. Then the distance to mid eclipse, $\eta$, is given by

$$\eta = \frac{z^3}{z^2 + s^2}.$$

Let the distance IH be equal to $\Delta$. This is an important quantity because it tells us whether or not an eclipse can be observed. If the Moon just grazes the Sun/shadow circle at the moment of greatest eclipse, then I is coincident with H and $\Delta$ is zero. A positive value of $\Delta$ therefore signifies that a lunar eclipse does occur, or that a solar eclipse both does occur and can be seen from the given geographical location. The value of $\Delta$ is given by

$$\Delta = \sqrt{\eta^2 - \frac{\eta[z^2 - (R_m + R_s)^2]}{z}}.$$

The Moon is at first contact at distance $AI = \eta - \Delta$, and at last contact when $AF = \eta + \Delta$.

The magnitude, $M$, of the eclipse can be found from the expression

$$M = \frac{R_m + R_o - \text{abs}\left(\dfrac{sz}{\sqrt{s^2 + z^2}}\right)}{2R_p},$$

where abs represents the absolute value of the expression within the brackets, $R_o$ is the radius of the Sun in the case of a solar eclipse or the radius of the shadow in the lunar eclipse, and $R_p$ is the radius of the Sun or the Moon in a solar or lunar eclipse, respectively. The value of $M$ is a number representing the fraction of the lunar diameter obscured by the Earth's shadow at mid lunar eclipse, or the fraction of the solar diameter obscured by the Moon at mid solar eclipse. In both cases the measurement is taken along the common diameter. It is possible for $M$ to be greater than one, for example in a total lunar eclipse.

## *Eclipses in AstroScript*

The circumstances of both lunar and solar eclipses may be calculated using the AstroScript command *compute_eclipses(⟨string⟩)*. The argument ⟨string⟩ may be set either to 'sun' or to 'moon' depending on the type of eclipse. For example, you could calculate solar eclipses with the single-line AstroScript program

    compute_eclipses(sun)

but on the disk with the book there is a more comprehensive program called *eclipses.txt* which you can use to investigate eclipses of either sort, and whose contents are as follows.

    {Search for and display lunar or solar eclipses}

    {Clear the screen and put up a title}
    display_title(+++ Lunar and solar eclipses +++)

    {Which type of eclipse?}
    set_string(s1,sun)
    ask_for_string(Find eclipses of the Moon or Sun,s1)

```
{Carry out the calculation according to type}
if (s1=moon)
   display_title(*** LUNAR ECLIPSES ***)
   compute_eclipses(s1)
end_if

if (s1=sun)
   display_title(*** SOLAR ECLIPSES ***)
   compute_eclipses(s1)
end_if

{Check for an illegal entry}
if (s1!sun)
   if (s1!moon)
      set_normal_text(12)
      new_line
      write_string(+++ not an allowed value)
      new_line
      new_line
      set_normal_text(7)
   end_if
end_if

{Another go?}
ask_for_repeat
```

This program first asks which kind of eclipse you wish to investigate, and sets the string variable s1 to your choice. Entering either sun or moon causes *compute_eclipses* to begin the calculation. Any other input is illegal, and the program tells you so.

The command *compute_eclipses* carries out a search of every lunation (month) beginning with the one in progress on the date which you enter, and continuing for as many months as you specify. The search identifies the occurrence or likely occurrence of an eclipse using the 'rules of eclipses' outlined earlier in the chapter. If one is found, you are asked whether you wish to calculate its circumstances. Answer *N* for no and the search continues. Answer *Y* and the program tests to see if the eclipse can actually be observed (using the formula for Δ – see above). If it can, the times of the start, mid eclipse, and end are found, together with other events such as the start of the umbral phase in a lunar eclipse. The magnitude of the eclipse is also calculated. Here is an example:

```
                    +++ Lunar and solar eclipses +++
                    --------------------------------

Find eclipses of the Moon or Sun ............. ? sun


                      *** SOLAR ECLIPSES ***
                      ----------------------

Please input the start date for the search:
Calendar date (d,m,y; BC neg) ................ ? 1,8,1999
Search for how many months ................... ? 3
```

```
Time zone (h ahead of UT; W negative) ........ ? 0
Daylight saving (h ahead of zone t) .......... ? 1
Geographic longitude (d,m,s) ................. ? 0 2 5.6700
Geographic latitude (d,m,s) .................. ? 52 10 12.0000
Height above mean sea level (m) .............. ? 20.0

Searching lunation in progress on  1-Aug-1999

Solar eclipse certain on local date .....  11-Aug-1999
Calculate the circumstances (Y or N) ......... ? Y

                  Local circumstances of the Solar Eclipse
                  ----------------------------------------

11-Aug-1999 in time zone: 0 with daylight saving: 1
Geographical longitude: 0 2 6, and latitude: 52 10 12
Local civil time: 11 21 2; UT: 10 21 2; ET/DT: 10 21 2

Event                Zone Date    Zone Time    Azimuth    Altitude
------------------------------------------------------------------


First contact        11-Aug-1999    10  3       118 17      38 32
Mid eclipse          11-Aug-1999    11 21       141  0      47 46
Last contact         11-Aug-1999    12 39       169 41      52 48


------------------------------------------------------------------


Magnitude ...................................    0.946

Display the eclipse (Y or N) ................. ? n

Another location (Y or N) ?  ................. ? N

Searching lunation in progress on 15-Sep-1999: no solar eclipse found
Searching lunation in progress on 14-Oct-1999: no solar eclipse found

Again (Y or N) ............................... ? n
```

You have the choice of seeing a moving sky display of the eclipse. Do not choose this option unless you have a fast enough computer as otherwise the response will be disappointingly slow. If you select the option of displaying the eclipse, you are asked for the start time (hours before mid eclipse), the duration of the moving display (hours), the interval between frames in minutes of eclipse time, the zoom factor for the display (1–50), and whether you wish to centre the display on the eclipse (moving window) or on a fixed point in the sky (fixed window). If you choose the moving window option, you are then asked whether the moving window is to be centred on the Sun or the Moon (solar eclipse), or the Moon or the Earth's umbra (lunar eclipse). You can also add an extra delay between frames, but unless your computer is super-fast you probably won't want to slow the display down. During the moving graphics display, press *I* to zoom in and *O* to zoom out. Press *B* to make the eclipse go backwards in time, or *F* to make it go forwards again. Press *S* then any key to step through the eclipse frame by frame (press *S* again to restart the motion). These are the only active keys. Pressing anything else will interrupt the display and allow you to return to the calculations.

# *AstroScript commands*

In this appendix, I describe the function of each AstroScript command in detail. I use a short-hand convention to signify the types of argument as follows:

⟨char⟩      This represents any single character **except** for '(', ')', and the delimiter (usually ',' but see the *set_delimiter* command below and *delimiter* in *Appendix B*). It can also represent either of the character variables as 'c1' or 'c2'. Thus ⟨*char*⟩ can be 'q' or 'T' or 'c2', but may not be ')' etc.

⟨char variable⟩      This represents only c1 or c2 and nothing else.

⟨string⟩      This represents any sequence of between 1 and 255 characters of any sort **except** for '(', ')', and the delimiter (usually ',' but see the *set_delimiter* command below and *delimiter* in *Appendix B*). It can also represent any of the three string variables as 's1' or 's2' or 's3'. Thus ⟨*string*⟩ can be '11 o'clock', 'Saturn' or 's3', but not 'Choose a value (3 or 4)' because this string contains two forbidden characters. You would need to rewrite it using one of the alternative braces e.g. 'Choose a value [3 or 4]'.

⟨string variable⟩      This represents only s1 or s2 or s3, and nothing else.

⟨prompt⟩      This represents the string which is displayed on the screen as a prompt for keyboard input. Exactly the same rules apply for ⟨*prompt*⟩ as for ⟨*string*⟩ above. Thus ⟨*prompt*⟩ could be 'What is the name of the planet'. Note that a prompt has a question mark (?) added automatically at the end, so you do not need to include one. It is also padded out with dots to make sure that keyboard input and program output are arranged neatly under each other on the screen. The above example would produce:

```
What is the name of the planet .............. ?
```

⟨accept⟩      This represents a text string of between 1 and 255 characters of any sort **except** for '(', ')', and the delimiter (usually ',' but see *set_delimiter* below and *delimiter* in *Appendix B*) which are the only characters allowed in the subsequent keyboard input. Thus if ⟨*accept*⟩ was set to 'yn', then only the characters 'Y' or 'N' (in either upper or lower case) would be accepted by the command from the keyboard.

AstroScript commands are grouped below into several categories according to their function.

## *General housekeeping commands*

*copy_to_printer(⟨string⟩)*
*copy_to_file(⟨string⟩)*

These commands cause most of what is displayed on the screen to be copied either to the printer or to a text file. The value of ⟨*string*⟩ must either be 'begin' to turn on the copying, or 'end' to turn it off. If not otherwise specified in the set-up file, *setup.dat* (see *Appendix B*) or using the *set_output_file_name* command (see below), the text file is given the name *output.prn*. If a file of this name already exists when the *copy_to_file* command runs, you are asked if you wish to write over it. Answer *Y* to delete the old file and begin afresh, or *N* to disregard the *copy_to_file* command. Examples are *copy_to_file(begin)* and *copy_to_printer(end)*. Graphics output, such as that used to display eclipses, is not copied.

*if (⟨char variable or string variable⟩⟨condition⟩⟨char or string⟩)*
*end_if*

The *if* . . . *end_if* structure enables you to write conditional clauses into your AstroScript program. The symbol ⟨*condition*⟩ represents one of the two characters '=', meaning 'is equal to', and '!', meaning 'is not equal to'. You may test only the values of the five variables c1, c2, s1, s2, or s3, and ⟨*char or string*⟩ may not be set to 'c1', 'c2', 's1', 's2', or 's3'. Thus *if (s1=saturn)* and *if (c1!.)* are both allowed, but *if (s2=s3)* is not. Each *if* command must be matched by exactly one *end_if* command, and all the AstroScript commands in between (any number from zero upwards) are processed only if the condition expressed by the argument is true. *If* . . . *end_if* structures may be nested to any level. For example, the following AstroScript fragment asks for the delay time only on the first pass through the program (when c1 is pre-set to '.'), but not on subsequent passes (when c1 has been set to a space):

```
        if (c1=.)
            ask_for_string(What delay time should be used,s1)
            set_character(c1, )
        end_if
```

Here is another example, with a nested 'if' block:

```
        if (c1=.)
            ask_for_character(Set c1 to a space [y or n],yn,c2)
            if (c2=y)
                write_string(c1 set to a space)
                new_line
                set_character(c1, )
            end_if
            if (c2!y)
                write_string(c1 left equal to a dot)
                new_line
            end_if
        end_if
```

The tests for equality and inequality are not sensitive to case. Thus, if s1 is set to 'mercury', both *if (s1=mercury)* and *if (s1=Mercury)* are true and produce the same result.

*repeat*
*ask_for_repeat*
*repeat_for_count(⟨string⟩)*

These commands give you the ability to repeat the whole program if you so wish. One of them can appear just once in your script program as the **last** command, although the interpreter will not produce an error if you put one elsewhere. The first form, *repeat*, simply repeats the whole program from the top without resetting any flags or changing any other variables. You can interrupt the program by pressing any key, or holding down the *cntrl* key and pressing *break*. The second form, *ask_for_repeat*, asks you if you wish to repeat the program as follows:

```
Again (Y or N) ............................... ? _
```

If you respond *N* (no) the program terminates normally. If you respond *Y* (yes) the program then asks if you wish to use new values for any of the variables you have previously typed in at the keyboard, or have set using a *set* command (see below), before rerunning the program from the first line. You will be asked only for new values of the variables you have selected, with the old values being used for the rest. The third form, *repeat_for_count*, is like the simple *repeat* command except that you can specify how many times the program must repeat before terminating normally. The argument, ⟨string⟩, must resolve to a positive integer number in the range 0 to 65535. Thus *repeat_for_count(s1)* is legal, and will behave in exactly the same way as *repeat_for_count(20)* if s1 has previously been set to '20'.

*wait_time_step(⟨string⟩)*

You can run your AstroScript program in 'real time' using this and the simple *repeat* command. The argument, ⟨string⟩, must resolve to a real number which is interpreted as an interval in seconds of time. The number is taken to be positive whatever its sign. The command suspends execution of the program until the computer's internal clock is equal to or greater than the last-used value of the local_civil_time plus the number of seconds specified by ⟨string⟩. The local_civil_time is then set to this new value before the program proceeds further. (The local_civil_time is a variable in the data bank, see Appendix C.) If the waiting interval spans midnight, the date is changed too. For example, this short script displays the solar and sidereal times every 5 seconds:

```
find_time(sidereal)
display_title(Times)
display_time(solar)
display_time(sidereal)
wait_time_step(5.0)
repeat
```

The command checks the computer's clock every integer second, so there is no difference between using wait steps of, say, 5.0 and 5.2, both causing a waiting interval of 5.0 seconds. Note that in real-time applications it is best to put the calculation part of the script before the display part. Here, *find_time(sidereal)* appears before *display_title(Times)* for example. This reduces flashing of the screen if the computer takes some time to make the calculation. The argument ⟨string⟩ must be a valid number representation made up of digits: '5.4', '20', and '2.34e1' are fine, but 'three' and '5s' are not. You can also specify the delay using s1, s2, or s3. For example, the following AstroScript fragment allows the delay to be specified first time through the program:

```
{s1 is preset to '.' so this if-block is executed only on the first pass}
if (s1=.)
   ask_for_string(Delay period [seconds],s1)
end_if

find_position(sun)
display_position(sun)

wait_time_step(s1)
repeat
```

*pause*
*pause(〈string〉)*
*suspend(〈string〉)*

These commands allow you to interrupt the program flow so that you can view intermediate results on the screen. The *pause* command may be used with or without an argument. When executed it displays 〈*string*〉 if present, or the message 'Press any key to continue . . .' if it isn't, and then waits until a key is pressed. The *suspend* command simply suspends the execution of the program for a period indicated by 〈*string*〉 in seconds. This argument must resolve to a real number in the range 0 to 65.535 seconds. For example, the command *suspend(32.81)* inserts a delay of 32.81 seconds.

*set_string(〈string variable〉, 〈string〉)*
*set_character(〈char variable〉, 〈char〉)*

The character variables, c1 and c2, and the string variables, s1, s2, and s3, are all preset to the dot or point character '.' when the AstroScript interpreter starts up. You can set any of these variables to any other legal values within your script program using the *set_character* or *set_string* commands. Thus, to set c1 equal to 'Y' you could use *set_character(c1,Y)*, or to set s3 equal to 'The Moon's position is' write *set_string(s3,The Moon's position is)* in your program.

*set_dynamic_offset(〈string〉)*

The positions of bodies like the Moon and the planets in our Solar System can be calculated using theories of their motions. Given any *dynamical time* in the future or the past, the position of the body can be found with a precision which depends only on the accuracy of the theory. The dynamical time is that smooth uniformly flowing time which we assume to exist, and which we can measure quite accurately nowadays using atomic clocks. However, we don't use dynamical time in our everyday lives, but rather *universal time* which is based on the rotation of the Earth. The conversion between universal time and dynamical time cannot be predicted with precision because the Earth's rotation is not uniform but exhibits small random changes from year to year. We do have historical records from which we can deduce the difference between dynamical and universal time in the past, and AstroScript incorporates algorithms for finding this difference for all years before 1988 (but with variable accuracy). You can set your own time difference by using *set_dynamic_offset*. The argument 〈*string*〉 must resolve to a real number (positive or negative) which represents the difference between dynamical time (DT) and universal time (UT) in hours i.e. DT − UT. Thus, to set this difference to −120 seconds you would use *set_dynamic_offset(−0.03333)*. If the *set_dynamic_offset* command appears in your AstroScript program, the given offset is used for all subsequent calculations irrespective of the date. If the command does not appear in your program, AstroScript uses its best endeavours to estimate the time difference for all dates before 1988, and sets the difference to zero for all dates after 1988. If you wish to turn off the dynamical time estimation, include the command *set_dynamic_offset(0)* at the beginning of your AstroScript program. Calculations will then assume that the given universal time (found from the local civil time, time zone, and daylight saving values) is also the dynamical time.

*set_delimiter(⟨char⟩)*
*set_decimal_places(⟨string⟩)*
*set_date_format(⟨string⟩)*
*set_normal_text(⟨string⟩)*
*set_inverse_text(⟨string⟩)*
*set_normal_background(⟨string⟩)*
*set_inverse_background(⟨string⟩)*
*set_aspect_ratio(⟨string⟩)*
*set_maximum_step(⟨string⟩)*
*set_number_of_iterations(⟨string⟩)*
*set_tolerance(⟨string⟩)*
*set_output_file_name(⟨string⟩)*

These 'set' commands give you control over some aspects of the 'look and feel' of AstroScript from inside the program. The default values for each parameter can be defined in the *setup.dat* file (see *Appendix B*) which is scanned once at the start of each program, but you can set these parameters to any other values in your program. In each case the argument must resolve to a legal value of the corresponding parameter, as defined in *Appendix B*. For example, change the colour of the displayed text using:

        ask_for_string(New text colour [0-15],s1)
        set_normal_text(s1)

Note the use of square brackets in the prompt string, since round brackets () are not allowed.

*set_longitude(⟨string⟩)*
*set_latitude(⟨string⟩)*
*set_height(⟨string⟩)*
*set_time_zone(⟨string⟩)*
*set_daylight_saving(⟨string⟩)*
*set_vertical_shift(⟨string⟩)*
*set_local_civil_time(⟨string⟩)*
*set_calendar_date(⟨string⟩)*

Whenever an AstroScript command needs the value of some parameter such as your geographical longitude which you have not yet specified, it automatically asks you for it. However, you can pre-empt this behaviour by using one of these 'set' commands. In each case, the argument ⟨string⟩ must resolve to a legal value of the corresponding parameter. The longitude, latitude, and vertical shift must be given in the format degrees, minutes, seconds, the time in the format hours, minutes, seconds, and the date in the format day, month, year (all digits; the order is independent of the *date_format* setting). Examples are *set_longitude(52,10,12.34)* which sets the geographical longitude to 52 degrees 10 minutes and 12.34 seconds north, *set_calendar_date(14,08,1994)* which changes the calendar date to 14th August 1994, *set_local_civil_time(23,21,0)* which sets the local civil time to twenty-one minutes past eleven pm, and *set_time_zone(−6.0)* which sets the time zone to −6 hours.

*increment_normal_text(⟨string⟩)*
*increment_inverse_text(⟨string⟩)*
*increment_normal_backgound(⟨string⟩)*
*increment_inverse_background(⟨string⟩)*

You can use these 'increment' commands to make relative changes to the values of the text-colour parameters inside your AstroScript program. In each case the argument ⟨string⟩ must resolve to an integer number (positive or negative) which is then added to the current value of the corresponding parameter. Legal values are given in *Appendix B*, but the program will attempt to interpret any integer value you supply, usually producing flashing coloured text. Try this example:

```
display_title()   {Clears the screen}
write_string(Have a nice )
increment_normal_text(1)
write_string(day!)
new_line
increment_normal_text(1)
suspend(1)
repeat
```

*increment_longitude(⟨string⟩)*
*increment_latitude(⟨string⟩)*
*increment_height(⟨string⟩)*
*increment_time_zone(⟨string⟩)*
*increment_local_civil_time(⟨string⟩)*
*increment_calendar_date(⟨string⟩)*

These 'increment' commands add the real number (positive or negative) indicated by the argument ⟨string⟩ to the corresponding variable. In the cases of longitude and latitude, the argument is in decimal degrees, in the case of height it is in decimal metres, in the cases of time zone and time it is in decimal hours, and in the case of date it is in days. Hence to add 30 minutes to the time you could use *increment_local_civil_time(0.5)*, and to change the date by 1 week you could write *increment_calendar_date(7)*.

## *Input commands*

*ask_for_character(⟨prompt⟩, ⟨accept⟩, ⟨char variable⟩)*

Use this command to set one of the character variables, c1 or c2, to the value you type at the keyboard. The command first displays the text string ⟨prompt⟩ beginning at the current cursor position, and then waits for you to type a character (followed by the *return* or *enter* key). It offers the current value of the character variable as the default value provided that it is one of the characters in the ⟨*accept*⟩ string. If it is not then the default value is the first character of the ⟨*accept*⟩ string. For example, if the character variable had been set to 'w', then the default value offered by the *ask_for_character* command would indeed be 'w' if ⟨*accept*⟩ was, say, 'qrtwe'. If however the character variable had the value 'x', then the default value offered would be 'q' since 'x' is not one of the allowed values and 'q' is the first in the list. When you respond, the character you type is checked against the ⟨*accept*⟩ list. If it is one of the characters in the list, the ⟨*char variable*⟩ is set to it and the program continues. Otherwise, the computer issues a warning beep and waits until you do type one of the allowed characters. Note that characters are matched without regard to their case, so that 'Y' matches 'y' in the accept string and vice versa. For example, the following accepts only 'Y', 'y', 'N', or 'n' in response to the question 'Had enough ......................?'

ask_for_character(Had enough,yn,c1)

When *return* or *enter* is pressed, the character variable c1 is set to the one of the four possible values which you typed. The padding of dots and the question mark are added automatically.

*ask_for_string(⟨prompt⟩, ⟨string variable⟩)*

You can use this command to set one of the string variables, s1, s2, or s3 to a string of characters that you type in. The ⟨*prompt*⟩ string is first displayed on the screen at the current cursor position with dot padding and question mark appended. The default value offered is the current value of the string variable. For example, you might use the following command in an AstroScript program to enter the planet whose position you wish to calculate.

ask_for_string(What is the name of the planet,s1)

*ask_for_coordinates(⟨char⟩)*

This command prompts for, and then accepts, pairs of coordinates from the keyboard according to the setting of ⟨*char*⟩. Valid settings are 'h' for hour angle/declination, 'r' for right ascension/declination, 'a' for azimuth/altitude, 'e' for ecliptic longitude/latitude, and 'g' for galactic longitude/latitude. The default values offered for the variables are their current settings. Thus *ask_for_coordinates(r)* would ask you to supply the right ascension and declination, as would *ask_for_coordinates(c1)*, if c1 had previously been set to 'r'. The argument can be in upper or lower case.

*clear_flags(⟨string⟩)*

This command resets the flags associated with data that have been entered via the keyboard, assigned using one of the 'set' commands, or produced in the course of making calculations. AstroScript asks for each item of information only once, raising a flag to signify that it has been placed in the data bank and using this previously entered value when it is needed later in the program, unless the *clear_flags* command has been run in between. The argument ⟨*string*⟩ can be set to 'all' to clear every flag in the data bank, or it can be set to 'date' to clear only those flags concerned with changes in the date and time.

*edit_star_data*

A text file (*stardata.asc*) of up to 2000 entries is used by the command *compute_sky* (see below) to hold the positions of stars and other objects. The command *edit_star_data* can be run to modify the file independently of *compute_sky*, for example to add new stars to the list. You can also use your text editor for the same purpose. Be cautious about making modifications. It is very easy to get in a muddle, so always make a copy of *stardata.asc* before you begin. Full details of the star file can be found in the section on *The star list*.

## *Calculation commands*

*convert_coordinates( ⟨char ⟩, ⟨char ⟩)*
*set_coordinates_type( ⟨string ⟩)*

The *convert_coordinates* command converts the coordinates from the system specified by the left-hand argument into the corresponding coordinates specified by the right-hand argument. Valid references for either argument are 'h' for hour angle/declination, 'r' for right ascension/declination, 'a' for azimuth/altitude, 'e' for ecliptic longitude/latitude, and 'g' for galactic longitude/latitude. An example is *convert_coordinates(r,g)* which takes the current values of the right ascension and declination and converts them into their corresponding galactic longitude and latitude. There are also special cases for the **left-hand** argument of 's', 'm', 'p', or 'o'. In each case the previously calculated right ascension and declination of the *S*un, *M*oon, a *p*lanet, or an *o*sculating position, respectively, are converted into the coordinates specified by the right-hand argument. (See under *find_position* for an explanation about osculating calculations.) By default, the topocentric coordinates are used for the Sun and the Moon, apparent geocentric coordinates for the planets, and true astrometric coordinates for the other bodies. However, you can specify otherwise by first using the auxiliary command *set_coordinates_type*. Its ⟨*string*⟩ argument must resolve to one of 'true', 'apparent', or 'topocentric'. All three types are available only for the Sun. In the other cases, the coordinates are selected according to Table 3.

For example, the following lines of AstroScript convert the Moon's topocentric right ascension and declination (calculated using *find_position(moon)*) into its topocentric azimuth and altitude, and display the result on the screen:

```
find_position(moon)
set_coordinates_type(topocentric)
convert_coordinates(m,a)
display_coordinates(a)
```

Table 3. *Coordinates selected by* convert_coordinates *for various arguments of* set_coordinates_type

| Argument | Sun | Moon | Planets | Osculating |
|---|---|---|---|---|
| 'true' | true geocentric coordinates | true geocentric coordinates + nutation | true geocentric coordinates | true astrometric coordinates |
| 'apparent' | as *true* + nutation + aberration | as *true* | as *true* + nutation + light time | as *true* |
| 'topocentric' | as *apparent* + parallax | as *true* + parallax | as *apparent* | as *true* |

*correct_for_precession*

This command performs a rigorous precession calculation, replacing the current values of right ascension and declination in the data bank with their precessed values.

*correct_for_parallax( ⟨string⟩)*

Use this command to correct for geocentric parallax. The argument ⟨*string*⟩ must resolve to 'topocentric to geocentric' or 'geocentric to topocentric' depending on which way round the calculation must be performed. The current values of the right ascension and declination in the data bank are replaced by their corrected values. An example is *correct_for_parallax(topocentric to geocentric)* which converts the right ascension and declination, assumed to be topocentric values, into the corresponding geocentric values.

*correct_for_refraction( ⟨string⟩)*

This command replaces the current values of right ascension and declination in the data bank with values corrected for atmospheric refraction. The direction of the correction is specified by ⟨*string*⟩ which must resolve either to 'apparent to true' or to 'true to apparent'. For example, these lines of AstroScript perform a correction of the right ascension and declination for atmospheric refraction, converting the apparent values as seen from the ground into the true values which would be observed if there were no atmosphere:

```
set_string(s1,apparent to true)
ask_for_coordinates(r)
correct_for_refraction(s1)
new_line
write_string(Coordinates corrected from )
write_string(s1)
write_string( values are:)
new_line
display_coordinates(r)
new_line
```

*find_time( ⟨string⟩)*

This command calculates the sidereal or solar times both for the Greenwich meridian (GST or UT) and for the local meridian (local sidereal time or local civil time). The argument ⟨*string*⟩ must resolve either to 'solar' or to 'sidereal'. Thus to calculate the sidereal time you could write *find_time(sidereal)*.

*find_date( ⟨string⟩)*

You can convert between the local calendar date and the local Julian date in either direction using this command. The corresponding dates on the Greenwich meridian are also calculated. The argument ⟨*string*⟩ must resolve either to 'calendar' or to 'julian'. For example, to find the calendar date corresponding to the current value of the Julian date in the data bank you could use *find_date(calendar)*.

*find_rise_set(⟨string⟩)*

This command calculates the circumstances of rising and setting of the object indicated by the value of ⟨*string*⟩. The argument must resolve to 'sun', 'moon', 'mercury', 'venus', 'mars', 'jupiter', 'saturn', 'uranus', or 'neptune' to perform the calculation for the corresponding celestial body. If the argument is none of these, then the calculation is made for the object whose right ascension and declination are currently held in the data bank, previously entered, for example, using *ask_for_coordinates(r)*.

*find_position(⟨string⟩)*

This command can be used to find the position in orbit of any of a number of celestial bodies, depending on the setting of ⟨*string*⟩. Recognised values are 'sun', 'moon', 'mercury', 'venus', 'mars', 'jupiter', 'saturn', 'uranus', or 'neptune', and in each case the position of the corresponding body is calculated using a theory describing its orbit around the Sun. Try, for example, the following lines of AstroScript which calculate and then display the position of Mars:

```
find_position(mars)
display_position(mars)
```

However, there are also many occasions, such as when calculating the position of a comet, when you do not have an exact theory for the orbit, but can only describe its motion over a relatively short interval in terms of an exact parabolic or elliptical orbit until the error between fact and calculation becomes too large to be useful. The elements which describe this exact orbit are called *osculating* elements because they need changing all the while to fit the real orbit. You can use the *find_position* command for these objects too by setting ⟨*string*⟩ to 'elliptical' or 'parabolic'. The command prompts you to supply the osculating elements themselves.

*find_when(⟨string⟩, ⟨string⟩, ⟨string⟩)*
*set_messages(⟨string⟩)*

The calculation commands described above find *what* happens at a given instant. For example, you supply the time and the date to *find_position(sun)*, and it returns with the Sun's position calculated for that instant. Yet often you want to make the calculation the other way around. You might wish to know *when* the altitude of the Sun is 10 degrees, or *when* the Moon passes through galactic longitude zero, or *when* Venus has a right ascension of 12.5 hours. The *find_when* command does all of this for you. The left-hand argument can be set to the name of a body, the centre argument to the name of a coordinate, and the right-hand argument to a target value.

   Valid references for the left-hand argument are 'sun', 'moon', 'mercury', 'venus', 'mars', 'jupiter', 'saturn', 'uranus', or 'neptune' to make the calculation for that particular body. If you set it to 'elliptical' or 'parabolic', the command asks you for the corresponding elements and makes an osculating orbit calculation. If the left-hand argument is set to anything else, the command assumes that you wish to make the calculation for the body whose right ascension and declination are those currently held in the data bank, which you may previously have entered using the *ask_for_coordinates(r)* command.

   Valid references for the middle argument are 'hour angle', 'right ascension', 'declination', 'azimuth', 'altitude', 'ecliptic longitude', 'ecliptic latitude', 'galactic longitude', and 'galactic latitude'. The right-hand argument must be a number, such as '12.5' or '237.2345' or '2.67e2'. (You must make sure that the target value is possible. You can't expect the command to find when the Sun's hour angle reaches 96!) For example, to find when the Sun's ecliptic longitude reaches 187.23 degrees you can write *find_when(sun,ecliptic longitude,187.23)*, or when a star called Deneb crosses the meridian (whose right ascension and declination you have already entered using *ask_for_coordinates(r)*) you would write *find_when(Deneb,azimuth,180)*, or equivalently *find_when(Deneb,hour angle,0)*.

Another use of the *find_when* command is to calculate the instant of the conjunction or opposition of two heavenly bodies, either in ecliptic longitude or in right ascension. In this case the middle argument must be set to 'conjunction in el', 'conjunction in ra', 'opposition in el', or 'opposition in ra' for conjunction in ecliptic longitude, conjunction in right ascension, opposition in ecliptic longitude, or opposition in right ascension, respectively. The right-hand argument then takes the name of the second body using the same rules as those described above for the left-hand argument. For example, to find the instant when the Moon and the Sun are next in opposition in right ascension you would use *find_when (moon,opposition in ra,sun)*, or when the planets Mars and Jupiter share the same ecliptic longitude you would write *find_when(jupiter,conjunction in el,mars)*. The left- and right-hand arguments must refer to different bodies.

This command performs an iterative search to converge on the solution. You must provide it with a starting point, and it then tries to find the answer closest to it. Give it another starting point to find another solution, say for another day (azimuth or altitude) or for another orbital cycle (right ascension etc.). The type of coordinates used by the command ('true', 'apparent', or 'topocentric') can be specified using the *set_coordinates _type* command (see *convert_coordinates* above). The progress of the iteration can be controlled to some extent using the *maximum_step*, *number_of_iterations*, and *tolerance* parameters described in *Appendix B*. They can also be adjusted using the corresponding *set* commands mentioned earlier.

The output of the *find_when* command can be turned on and off using *set_messages(⟨string⟩)*. Set ⟨string⟩ to *OFF* to suppress the output, useful when creating a data file with *copy_to_file* and *display_data*. Set ⟨string⟩ to *ON* to reinstate the output if you have previously turned it off.

## Display commands

*new_line*

This command moves the cursor down to the beginning of the next line.

*write_string(⟨string⟩)*

Running this command displays the text ⟨string⟩ on the screen at the current cursor position without adding a new-line character at the end. The length of ⟨string⟩ can be anything from 1 to 255 characters, but if the string is longer than the width of the screen it is wrapped around onto the next line. (Note, however, that you are limited to a shorter string length than 255 characters when using the AstroScript command input facility.) For example, the following AstroScript segment would write the message 'Have a nice day!', leaving the cursor at the beginning of the next line:

```
write_string(Have a )
write_string(nice day!)
new_line
```

*display_title(⟨string⟩)*

This command first clears the screen, and then writes the text ⟨*string*⟩ centred and underlined on the top line, leaving the cursor two lines below it. For example, you might wish to use *display_title(+++ Coordinate Conversions +++)* as the first line in an AstroScript program which converts coordinates from one system to another.

*display_date(⟨string⟩)*

This command displays the current calendar or Julian date held in the data bank. You can convert dates from one type to another using the *find_date* command, but it is often not necessary to run that command explicitly since the calculations may be made in the course of other commands. The argument ⟨*string*⟩ must resolve either to 'calendar' or to 'julian' to display the corresponding date on the screen. For example, in the following AstroScript program fragment, the calendar date is needed, and is asked for, by the *find_position* command. The calculation of the Julian date is also made in the course of the *find_position* command:

        find_position(mars)
        display_date(calendar)
        display_date(julian)

The following fragment does not work (try it for yourself)

        input_coordinates(r)
        display_date(julian)

since the Julian date has not been calculated before the display command.

*display_time(⟨string⟩)*

Use this command to display the local civil time and universal time, or the local sidereal time and the Greenwich sidereal time. The argument ⟨*string*⟩ must resolve to 'solar' for local civil and universal times, or to 'sidereal' to display the sidereal times. Try, for example, this short AstroScript program:

        display_title(Sidereal Time)
        find_time(sidereal)
        display_time(sidereal)
        ask_for_repeat

*display_time_difference(⟨string⟩, ⟨string⟩)*

This command is useful if you wish to display the difference between two times with the result expressed in h,m,s format. The two arguments can each represent the times to be subtracted from each other in decimal hours, and the value of the right-hand string is then subtracted from that of the left-hand string. Thus if the left-hand string was set to '11.5' and the right-hand string to '0.25', the command would display the difference $11.5 - 0.25$ hours in h,m,s format i.e. 11 15 0. Either string may also be set to 'local civil time', 'universal time', 'local sidereal time', or 'greenwich sidereal time', and the corresponding value is then extracted from the data bank and used in the command. For example, the following

AstroScript fragment displays the time difference between the local civil time and UT, and then the interval of local time since midday:

```
write_string(LCT − UT: )
display_time_difference(local civil time,universal time)
new_line
write_string(Time since noon: )
display_time_difference(local civil time,12.0)
new_line
```

*display_coordinates( ⟨char ⟩)*

This command displays the current values held in the data bank of the pair of coordinates referred to by ⟨*char* ⟩. Valid references are 'h' for hour angle/declination, 'r' for right ascension/declination, 'a' for azimuth/altitude, 'e' for ecliptic longitude/latitude, and 'g' for galactic longitude/latitude. There are also four special cases of 's', 'm', 'p', and 'o', referring to the right ascension and declination of the *S*un, *M*oon, a *p*lanet, or the result of an *o*sculating orbit calculation respectively. You can specify whether true, apparent or topocentric coordinates are to be displayed using the command *set_coordinates_type* (see under *convert_coordinates* for details and restrictions). For example, the following will display the Moons topocentric right ascension and declination:

```
find_position(moon)
set_coordinates_type(topocentric)
display_coordinates(m)
```

*display_data( ⟨string ⟩, ⟨string ⟩)*

Data that you enter via the keyboard, and the results of the calculations, are stored in a common data bank which is available to every AstroScript command. Normally you display these on the screen using one of the display commands such as *display_position* or *display_rise_set*. However, it is sometimes useful to examine individual items in the data bank, or to create a disk file containing certain values only, and the *display_data* command lets you do just that. It has two arguments. The first (left-hand) argument can be any string of characters (other than '(', ')', or the delimiter (usually ',' but see the *set_delimiter* command above and *delimiter* in *Appendix B*)) which is displayed on the screen padded with dots to make it the standard length. This can be used to describe the result being displayed. The dot padding is not added if the left-hand string contains no characters. Instead, the empty string is replaced by a single space. This facility allows you to create a list of values suitable for exporting into other programs such as a spread sheet. (See the second example below.) The second (right-hand) argument must resolve to the name of the item in the data bank that you wish to display. There is a list of all the items in *Appendix C*. Notice that many of them are *flags* which are either 'true' or 'false'. They show, for example, whether particular groups of items have been calculated yet, being set to 'false' if not. Note that the command does not finish with a new-line character, so that the cursor remains on the same line after displaying the data. Use the *new_line* command if you wish the next piece of output to appear at the beginning of the next line. Consider the following fragment of AstroScript:

```
display_data(Moon geocentric ecliptic longitude is,moon_long)
new_line
display_data(Moon ecliptic coordinates are valid is,moon_ecl_valid)
```

```
new_line
new_line
find_position(moon)
new_line
display_data(Moon geocentric ecliptic longitude now is,moon_long)
new_line
display_data(Moon ecliptic coordinates are valid is,moon_ecl_valid)
new_line
```

When I ran the program it produced the following output:

```
Moon geocentric ecliptic longitude is ........   ** unset **
Moon ecliptic coordinates are valid is .......   ** false **

Calendar date (d,m,y; BC neg) ................ ? 9/ 3/ 1995
Local civil time (h,m,s) ..................... ? 20 31 32.0000
Time zone (h ahead of UT; W negative) ........ ? 0
Daylight saving (h ahead of zone t) .......... ? 0
Geographic longitude (d,m,s) ................. ? 0 2 5.6700
Geographic latitude  (d,m,s) ................. ? 52 10 12.0000
Height above mean sea level (m) .............. ? 20.0

Moon geocentric ecliptic longitude now is .... 83.460279
Moon ecliptic coordinates are valid is ....... ** true **
```

The data that I entered via the keyboard are underlined. You can see that there were no valid data to display before the *find_position(moon)* command, so the first two lines produced the messages ** *unset* ** and ** *false* ** in response to the *display_data* commands. When the Moon's position had been calculated, however, the *display_data* commands produced satisfactory results. Note that if you try to display the value of a data item not shown in *Appendix C*, perhaps because you made a mistake in your typing, the computer emits a warning beep and you get the message ** *not found* **.

The *display_data* command can also be used to produce a table of values suitable for exporting into another program, such as a spread sheet. In the following example, the disk file *table.prn* is created which contains the local civil time and the altitude of the Sun tabulated every hour for ten hours from the designated starting time:

```
set_output_file_name(table.prn)
set_delimiter(%)
find_position(sun)
copy_to_file(begin)
convert_coordinates(s%a)
display_data(%local_civil_time)
write_string(,)
display_data(%altitude)
new_line
increment_local_civil_time(1)
repeat_for_count(10)
```

The first command defines the name of the disk file to be created, in this case *table.prn*. The second command sets the delimiter character to '%' so that the comma character can be used by *write_string* further on in the program to separate the data in each line of the file. The position of the Sun is calculated by the third command, which asks for all the details it needs like the time and date. The fourth command turns on the copying; all output from now on is both displayed on the screen and is copied into *table.prn*. The next command converts the Sun's coordinates into azimuth and altitude. Commands six, seven, and eight write the local civil time, a comma, and the Sun's altitude on a single line. The opening bracket of the argument of each *display_data* command is followed immediately by the delimiter (%) so that the first argument has no characters in it. Hence dot padding is turned off. The *new_line* command then moves the cursor to the beginning of the next line down. The penultimate command adds one hour to the local civil time, and the last command causes the whole program to be repeated from the top ten times over. This is what was produced on the screen when I ran the program:

```
Calendar date (d,m,y; BC neg) ................ ? 9/ 3/ 1995
Local civil time (h,m,s) ..................... ? 0,0,0
Time zone (h ahead of UT; W negative) ........ ? 0
Daylight saving (h ahead of zone t) .......... ? 0
Geographic longitude (d,m,s) ................. ? 0 2 5.6700
Geographic latitude  (d,m,s) ................. ? 52 10 12.0000
Height above mean sea level (m) .............. ? 20.0
    0.000000, -42.543824
    1.000000, -41.490921
    2.000000, -37.455377
    3.000000, -31.115707
    4.000000, -23.240218
    5.000000, -14.472317
    6.000000, -5.312610
    7.000000, 3.822821
    8.000000, 12.534576
    9.000000, 20.380095
   10.000000, 26.833050
```

As before, the underlined values were entered by me from the keyboard. The disk file *table.prn* contained the following:

```
    0.000000, -42.543824
    1.000000, -41.490921
    2.000000, -37.455377
    3.000000, -31.115707
    4.000000, -23.240218
    5.000000, -14.472317
    6.000000, -5.312610
    7.000000, 3.822821
    8.000000, 12.534576
    9.000000, 20.380095
   10.000000, 26.833050
```

*display_rise_set( ⟨string⟩)*

Having made a calculation to find the circumstances of rising and setting, use this command to display the result. The argument ⟨*string*⟩ must resolve to 'sun', 'moon', 'mercury', 'venus', 'mars', 'jupiter', 'saturn', 'uranus', or 'neptune' to display the results for those particular bodies. If ⟨*string*⟩ resolves to anything else, the command displays the results of the calculation based upon the right ascension and declination in the data bank, entered perhaps by the *ask_for_coordinates(r)* command. Thus a program to find the times of rising and setting of venus could be

```
find_rise_set(venus)
display_rise_set(venus)
```

Note that you cannot put these two commands in the reverse order. If you try to display the rising and setting times before you have calculated them, you will get an error message!

*display_position(⟨string⟩)*

This command displays the results of orbital calculations made using the *find_position* command. Set ⟨*string*⟩ to 'sun', 'moon', 'mercury', 'venus', 'mars', 'jupiter', 'saturn', 'uranus', 'neptune', 'elliptical', or 'parabolic', as for *find_position*. For example, the following two lines of AstroScript are all you need to display the position of Jupiter:

> find_position(jupiter)
> display_position(jupiter)

Note that you cannot display the position of a planet different from the one whose position you have just calculated. If the second line above had read, say, *display_position(mars)*, then the computer would have responded with an error message.

## *Complex commands*

*compute_nutation*
*compute_obliquity(⟨string⟩)*

Use the command *compute_nutation* to calculate and display the values of nutation in longitude and nutation in obliquity, or the command *compute_obliquity* to display the value of the obliquity of the ecliptic. The obliquity may be displayed with or without a correction for nutation according to the setting of the argument ⟨*string*⟩, which must resolve either to 'include nutation' or to 'exclude nutation'. For example, to display the obliquity without including nutation you would write *compute_obliquity(exclude nutation)*.

*compute_moon_phases*

Run this command to find the instances, correct to a few seconds, of the phases of the moon during the lunation in progress on the given date. The command also detects and displays the likely occurrence of a lunar or solar eclipse (see also *compute_eclipses* below).

*compute_ephemeris(⟨string⟩)*

The *compute_ephemeris* command allows you to tabulate the position of the body specified by ⟨*string*⟩ over a given period at selected intervals. For example, you could use *compute_ephemeris(moon)* to produce a table of the Moon's topocentric position for every hour over a 10-hour period starting at a given date and time. The argument ⟨*string*⟩ must resolve to 'sun', 'moon', 'mercury', 'venus', 'mars', 'jupiter', 'saturn', 'uranus', or 'neptune'. It is not possible to produce an ephemeris for Pluto, or any other osculating orbit by this means.

*compute_eclipses(⟨string⟩)*

This is a very powerful command, and it constitutes a whole suite of programs in a single line. It performs a search over a specified interval beginning at a given date for the occurrence of a solar or lunar eclipse depending on whether ⟨*string*⟩ resolves to 'sun' or 'moon', respectively. If an eclipse is detected, its details can be calculated. There is also the option of viewing the eclipse as a moving graphical image of the sky (if your computer is fast enough!). You can zoom in or out, move forwards or backwards in time, or step through the sequence one step at a time. See the section called *Eclipses* for more details.

*compute_sky*

This is the most powerful command of all, and it incorporates most of the actions of the other commands within a single line of program. You are presented with four pages of information set out in tabular form. Page 1 gives the positions of the Sun, Moon, and seven major planets, either in real time or at another specified time. Page 2 displays the rising and setting circumstances of these same objects. Pages 3 and 4 display the positions and rising/setting circumstances, respectively, of all the objects whose details are contained in the star data file, which you can edit within this command, or separately using the *edit_star_data* command. You can also obtain a graphical image of the sky, provided that your computer is quick enough. See the section called *Sky graphics* for more details.

# *The configuration file setup.dat*

Some aspects of the look and feel of AstroScript, such as the text colours and the number of decimal places displayed, can be defined in a text file called *setup.dat* which is read every time AstroScript starts up. Here you may also specify some of the default values used by the interpreter when running your program, including your geographical position. The file distributed on the disk which comes with the book has my own settings in it. You can edit this as you like.

Each command within the file sets the default value for just one parameter. The name of the parameter comes first, then a space, and then the value to which it is to be set. Comment lines can be included by beginning each such line with the '{' character, and there can also be any number of blank lines. Any or all of the following parameters can be included in the file in any order. Those not included are given default settings specified within the *ascript.exe* program itself. Note also that you can set any parameter yourself within an AstroScript program using the *set* commands (see *Appendix A*).

*delimiter ⟨char⟩*

Here ⟨*char*⟩ represents any character. This line defines the character used to separate arguments in AstroScript commands having two or more arguments. By default I use ',' the comma character, but you can change this to anything else you like. For example, if the *setup.dat* file contains the line *delimiter %*, then you would have to use '%' in all multiple-argument commands (unless changed in a program using *set_delimiter* – see *Appendix A*), e.g. *convert_coordinates(m%a)* instead of *convert_coordinates(m,a)*. Note that wherever you see the ',' symbol in this book used within the brackets containing the arguments of an AstroScript command, you can replace it by whatever you specify in the *setup.dat* file. This facility gives you the flexibility to use any particular character in a string argument which would otherwise be interpreted as a delimiter. (See, for example, the AstroScript program in the section called *Converting coordinates using AstroScript.*)

*decimal_places ⟨n⟩*

This line defines the number of places of decimals to be used whenever a decimal number is displayed on the screen. Here ⟨*n*⟩ represents an integer value from 0 to 7. The default is *decimal_places 4*. Specifying a number greater than 7 is the same as writing *decimal_places 7*, and similarly a negative number produces the same result as *decimal_places 0*. The value of ⟨*n*⟩ does **not** define the accuracy of the result. You can display as many decimal places as you like (within the limits given above), but often only the first few digits might be meaningful.

*aspect_ratio ⟨d⟩*

Some computer screens, especially the LCD screens used in laptop models, have different aspect ratios from those of the standard desktop variety. Circles then appear as ellipses. By setting ⟨*d*⟩ equal to the appropriate decimal multiplier, you can ensure that the circles do indeed appear to be round on your screen. For standard screens set *aspect_ratio 1.00* (default value); for laptop screens you may need to set *aspect_ratio 1.25*. This parameter can be set to any value, including less than or equal to zero, but with silly results.

*normal_text ⟨n1⟩*
*inverse_text ⟨n2⟩*
*normal_background ⟨n3⟩*
*inverse_background ⟨n4⟩*

These four lines define the colours used for displaying text. Here ⟨*n1*⟩ and ⟨*n2*⟩ are integers in the range 0 to 15, setting the colour of normal text foreground and inverse text foreground respectively. The values of ⟨*n3*⟩ and ⟨*n4*⟩ can be in the range 0 to 7, setting normal text background, and inverse text background respectively. The default settings are *normal_text 7*, *inverse_text 0*, *normal_background 0*, and *inverse_background 7*, representing light grey on black for normal text, and black on light grey for inverse text. The numbers represent foreground colours as shown in Table 4.

You may also invoke the blink attribute for the foreground colours by adding 128 to the above numbers. For example, the following sets the normal text colour to light blue on a blue background, and the inverse text to blinking white on a red background:

```
normal_text 9
inverse_text 143
normal_background 1
inverse_background 4
```

The eight background colours are the first eight in Table 4.

*time_zone ⟨d⟩*
*daylight_saving ⟨n⟩*

Your time zone and daylight saving corrections may be set using these lines. The value of ⟨*d*⟩ represents the number of hours that your time zone is ahead of the Greenwich zone. For example, you will need to set *time_zone 6.5* if your time zone is 6 hours and 30 minutes ahead. Listen to a broadcast from the BBC World Service to find the time at Greenwich

Table 4. *Foreground colours for displaying text. The first eight are also background colours*

| black | 0 | dark grey | 8 |
|---|---|---|---|
| blue | 1 | light blue | 9 |
| green | 2 | light green | 10 |
| cyan | 3 | light cyan | 11 |
| red | 4 | light red | 12 |
| magenta | 5 | light magenta | 13 |
| brown | 6 | yellow | 14 |
| light grey | 7 | white | 15 |

(given as Greenwich mean time, which you should understand to be UTC) and compare it with the time on your watch if you are not sure what setting to use. The integer ⟨*n*⟩ defines how many hours ahead is your local time compared to your zone time. Many countries add a correction of 1 hour during the summer months to make the daylight hours fit more comfortably into the civil day. This is often called the daylight-saving correction. Thus, if your local time has been advanced by 1 hour, set *daylight_saving 1*. Note that these two values are always used together in AstroScript calculations, so setting your *time_zone* to 5 and the *daylight_saving* to 1 produces the same result as setting *time_zone* to 6 and *daylight_saving* to zero.

*date_format ⟨string⟩*

The formats of calendar dates commonly used in the United Kingdom and in the United States are different. In the UK, it is usual to write the date in the order day/month/year, whilst in the US the order often used is month/day/year. You can instruct AstroScript to use either format by using the *date_format* command with the argument ⟨*string*⟩ set either to 'UK' or to 'US'. Thus to use the UK format (the default setting), write *date_format UK*, or for the US format write *date_format US*.

*longitude ⟨d,m,s⟩*
*latitude ⟨d,m,s⟩*

Set your default geographical longitude and latitude using these two lines, using the format degrees, minutes, seconds. Longitudes **east** of Greenwich are **positive** and those west of Greenwich are negative. Your latitude is positive if north of the equator, and negative south. Thus if you live in Buenos Aires (latitude 34° 40′ S, longitude 58° 30′ W) you might wish to incorporate the lines *longitude −58,30,0* and *latitude −34,40,0* in *setup.dat*.

*height ⟨d⟩*

This line specifies your height in metres above the standard spheroid (sea level will do since it is not a very sensitive parameter). For example, *height 30* specifies the default height of 30 metres.

*vertical_shift ⟨d,m,s⟩*

The effects of atmospheric refraction and/or hills on the horizon on the times of rising and setting can be taken into account using this parameter. See the section called *Atmospheric refraction* for more details. The vertical shift should be given in degrees, minutes, and seconds format. For a standard atmosphere and unobstructed horizon the value is 34 minutes of arc, so you should write *vertical_shift 0,34,0*. The value of *vertical_shift* is positive if the effect of the shift is to increase the length of the day and negative if the effect is to decrease the length of the day. Hence the effect of distant hills on your horizon could be taken into account by using a negative value of this parameter.

*maximum_step ⟨d1⟩*
*number_of_iterations ⟨n⟩*
*tolerance ⟨d2⟩*

The *find_when* command (see *Appendix A*) uses an iterative technique to calculate the time at which a particular event occurs, such as *when* the Sun crosses the meridian (azimuth = 180 degrees or hour angle = 0 hours). The value of *maximum_step* limits the largest step allowed in the iterative process. By default, ⟨*d1*⟩ has the value of 10, corresponding to a maximum change of plus or minus 10 days in the date for which the next test calculation is made. The value of *number_of_iterations* sets an upper limit to the number of steps that are allowed before the program gives up and reports that it could not find a solution. By default, ⟨*n*⟩ is set to 100. The value given to *tolerance* defines how close the solution must be to the correct value before the program returns the result. Thus if ⟨*d2*⟩ is set to 1.0, the program stops running when it has found a solution within one degree (or 1 hour if right ascension or hour angle) of the right answer. The default is 1.0e−6 (one millionth).

*output_file_name* ⟨*filename*⟩

The AstroScript command *copy_to_file* (see *Appendix A*) copies most things displayed on the computer screen to a text file. The default name is *output.prn*, but you can set ⟨*filename*⟩ to any other valid file name you like. Thus to send output to 'myfile.txt' you would write *output_file_name myfile.txt*.

# *Variables held in the data bank*

The values of the following variables can be displayed on the screen using the *display_data* command described in *Appendix A*:

| | |
|---|---|
| day_of_week | {Sunday = 0, Monday = 1 etc.} |
| dow_valid | {True if day_of_wcek is valid} |
| | |
| day | {Day number of the month, 1–31} |
| month | {Month number of the year, 1–12} |
| year | {Year number, BC negative, e.g. 1992, −438} |
| cal_date_valid | {True if day, month, year is valid} |
| | |
| g_day | {Day number of the month at Greenwich, 1–31} |
| g_month | {Month number of the year at Greenwich, 1–12} |
| g_year | {Year number at Greenwich, BC negative, e.g. −438} |
| g_date_valid | {True if the Greenwich calendar date is valid } |
| | |
| julian_date | {Julian day number: days since 1.5/Jan/4713 BC} |
| dynamic_julian_date | {Julian day number using dynamic time, days} |
| jul_date_valid | {True if the Julian day number is valid} |
| | |
| universal_time | {Universal coordinated time, hours} |
| dynamic_time | {Ephemeris/dynamic time estimate, hours} |
| ut_valid | {True if universal_time is valid} |
| | |
| local_civil_time | {Local civil time in hours} |
| lct_valid | {True if local_civil_time is valid} |
| | |
| universal_time1 | {Alternative UT if universal_time is ambiguous} |
| ut1_valid | {True if universal_time1 is valid} |
| | |
| local_civil_time1 | {Alternative LCT if local_civil_time is ambiguous} |
| lct1_valid | {True if local_civil_time1 is valid} |
| | |
| g_sid_time | {Sidereal time at Greenwich, longitude 0, hours} |
| gst_valid | {True if g_sid_time is valid} |
| | |
| local_sid_time | {Sidereal time on local meridian, hours} |
| lst_valid | {True if local_sid_time is valid} |
| | |
| time_zone | {Time zone of the observer, hours} |
| daylight_saving | {Hours ahead of zone time for daylight saving} |
| tz_valid | {True if time_zone and daylight_saving are valid} |

g_date_offset          {Greenwich date minus local date, days}

geo_longitude          {Geographical longitude, degrees}
geo_long_valid         {True if geo_longitude is valid}

geo_latitude           {Geographical latitude, degrees}
geo_lat_valid          {True if geo_latitude is valid}

geo_height             {Height above geoid (sea level good enough), metres}
geo_ht_valid           {True if geo_height is valid}

nutation_in_obl        {Value of nutation in the obliquity, degrees}
nutation_in_long       {Value of nutation in ecliptic longitude, degrees}
nutation_valid         {True if nutations are valid}

obliquity              {Value of the obliquity of the ecliptic, degrees}
obliquity_valid        {True if the value of obliquity is valid}

include_nutation       {True if nutation in obliquity is added to the obliquity}

hour_angle             {Hour angle, hours}
declination            {Declination, degrees}
hadec_valid            {True if hour_angle and declination are valid}

right_ascension        {Right ascension, hours}
radec_valid            {True if right_ascension and declination are valid}

azimuth                {Azimuth, degrees}
altitude               {Altitude, degrees}
azalt_valid            {True if azimuth and altitude are valid}

ecl_longitude          {Ecliptic longitude, degrees}
ecl_latitude           {Ecliptic latitude, degrees}
ecliptic_valid         {True if the ecliptic coordinates are valid}

gal_longitude          {Galactic longitude, degrees}
gal_latitude           {Galactic latitude, degrees}
galactic_valid         {True if the galactic coordinates are valid}

azha_mat_valid         {True if the az/alt ha/dec conversion matrix is valid}
raha_mat_valid         {True if the ra/dec ha/dec matrix is valid}
raecl_mat_valid        {True if the ra/dec ecliptic matrix is valid}
ragal_mat_valid        {True if the ra/dec galactic matrix is valid}

prec_daya              {Precession epoch: from day number...}
prec_montha            {...month number...}
prec_yeara             {...year number}
prec_dayb              {Precession epoch: to day number...}
prec_monthb            {...month number...}
prec_yearb             {...year number}

prec_mat_valid       {True if the precession matrix is valid}

temperature       {Atmospheric temperature for refraction model, °C}
pressure       {Atmospheric pressure for refraction model, mbar}
atmosphere_valid       {True if temperature & pressure have been set}

equ_hor_parallax       {Equatorial horizontal parallax, degrees}
ehp_valid       {True if equ_hor_parallax is valid}

rho_cos_phi       {$\rho \cos \phi'$, equatorial Earth radii (see *Geocentric parallax*)}
rho_sin_phi       {$\rho \sin \phi'$, equatorial Earth radii (see *Geocentric parallax*)}
r_dist       {Distance from Earth centre, equatorial Earth radii}
parallax_valid       {True if these parallax values are valid}

vertical_shift       {Vertical displacement at horizon, degrees}
vert_shift_valid       {True if vertical_shift is valid}

lst_rise       {Local sidereal time of rising, hours}
lst_set       {Local sidereal time of setting, hours}
gst_rise       {Greenwich sidereal time of rising, hours}
gst_set       {Greenwich sidereal time of setting, hours}
lct_rise       {Local civil time of rising, hours}
lct_set       {Local civil time of setting, hours}
ut_rise       {Universal time of rising, hours}
ut_set       {Universal time of setting, hours}
lct1_rise       {Alternative local civil time of rising if ambiguous, hours}
lct1_set       {Alternative local civil time of setting if ambiguous, hours}
ut1_rise       {Alternative universal time of rising if ambiguous, hours}
ut1_set       {Alternative universal time of setting if ambiguous, hours}
azimuth_rise       {Azimuth of rising, degrees}
azimuth_set       {Azimuth of setting, degrees}
rise_set_valid       {True if the rising and setting values are valid}

sun_long       {Sun's true geocentric ecliptic longitude, degrees}
sun_lat       {Sun's true geocentric ecliptic latitude, degrees}
sun_rv       {Sun's true radius vector, au}
sun_app_long       {Sun's apparent geocentric ecliptic longitude, degrees}
sun_ecl_valid       {True if the Sun's ecliptic coordinates are valid}

sun_ra       {Sun's true right ascension, hours}
sun_dec       {Sun's true declination, degrees}
sun_app_ra       {Sun's apparent right ascension, hours}
sun_app_dec       {Sun's apparent declination, degrees}
sun_top_ra       {Sun's topocentric right ascension, hours}
sun_top_dec       {Sun's topocentric declination, degrees}
sun_top_long       {Sun's topocentric ecliptic longitude, degrees}
sun_top_lat       {Sun's topocentric ecliptic latitude, degrees}
sun_radec_valid       {True if Sun's equatorial coordinates are valid}

lst_sun_rise       {Local sidereal time of sunrise, hours}
lst_sun_set       {Local sidereal time of sunset, hours}

| gst_sun_rise | {Greenwich sidereal time of sunrise, hours} |
|---|---|
| gst_sun_set | {Greenwich sidereal time of sunset, hours} |
| lct_sun_rise | {Local civil time of sunrise, hours} |
| lct_sun_set | {Local civil time of sunset, hours} |
| ut_sun_rise | {Universal time of sunrise, hours} |
| ut_sun_set | {Universal time of sunset, hours} |
| azimuth_sun_rise | {Azimuth of sunrise, degrees} |
| azimuth_sun_set | {Azimuth of sunset, degrees} |
| sun_rs_valid | {True if Sun's rising and setting values are valid} |
| | |
| lst_planet_rise | {Local sidereal time of planet rise, hours} |
| lst_planet_set | {Local sidereal time of planet set, hours} |
| gst_planet_rise | {Greenwich sidereal time of planet rise, hours} |
| gst_planet_set | {Greenwich sidereal time of planet set, hours} |
| lct_planet_rise | {Local civil time of planet rise, hours} |
| lct_planet_set | {Local civil time of planet set, hours} |
| ut_planet_rise | {Universal time of planet rise, hours} |
| ut_planet_set | {Universal time of planet set, hours} |
| azimuth_planet_rise | {Azimuth of planet rise, degrees} |
| azimuth_planet_set | {Azimuth of planet set, degrees} |
| planet_rs_valid | {True if planet's rising and setting values are valid} |
| | |
| earth_helio_long | {Earth's heliocentric longitude, degrees} |
| earth_helio_lat | {Earth's heliocentric latitude, degrees} |
| earth_rv | {Earth's radius vector, au} |
| planet_helio_long | {Planet's heliocentric longitude, degrees} |
| planet_helio_lat | {Planet's heliocentric latitude, degrees} |
| | |
| planet_long | {Planet's true geocentric ecliptic longitude, degrees} |
| planet_lat | {Planet's true geocentric ecliptic latitude, degrees} |
| planet_rv | {Planet's true radius vector, au} |
| planet_dist | {Planet's true distance from the Earth, au} |
| planet_app_long | {Planet's apparent geocentric ecliptic longitude, degrees} |
| planet_app_lat | {Planet's apparent geocentric ecliptic latitude, degrees} |
| planet_ecl_valid | {True if the planet's ecliptic coordinates are valid} |
| | |
| planet_ra | {Planet's true right ascension, hours} |
| planet_dec | {Planet's true declination, degrees} |
| planet_app_ra | {Planet's apparent right ascension, hours} |
| planet_app_dec | {Planet's apparent declination, degrees} |
| plan_radec_valid | {True if planet's equatorial coordinates are valid} |
| | |
| planet_elongation | {Solar elongation of a planet, degrees} |
| planet_ang_size | {Planet's angular diameter, degrees} |
| planet_phase | {Planet's phase, 0 = 'new', 1 = 'full'} |
| planet_pos_angle | {Position angle of planet's bright limb, degrees} |
| planet_par_angle | {Planet's parallactic angle, degrees} |
| plan_asp_valid | {True if the above five planetary aspects are valid} |
| | |
| moon_long | {Moon's apparent geocentric ecliptic longitude, degrees} |
| moon_lat | {Moon's apparent geocentric ecliptic latitude, degrees} |

| | |
|---|---|
| moon_dist | {Earth–Moon distance, km} |
| moon_parallax | {Horizontal equatorial parallax, degrees} |
| moon_ecl_valid | {True if the Moon's ecliptic coordinates are valid} |
| | |
| moon_ra | {Moon's apparent geocentric right ascension, hours} |
| moon_dec | {Moon's apparent geocentric declination, degrees} |
| moon_hour_angle | {Moon's apparent geocentric hour angle, hours} |
| moon_app_long | {Moon's topocentric ecliptic longitude, degrees} |
| moon_app_lat | {Moon's topocentric ecliptic latitude, degrees} |
| moon_app_ra | {Moon's topocentric right ascension, hours} |
| moon_app_dec | {Moon's topocentric declination, degrees} |
| moon_pos_valid | {True if the Moon's coordinates are valid} |
| | |
| moon_ang_size | {Moon's geocentric angular diameter, degrees} |
| moon_app_ang_size | {Moon's topocentric angular diameter, degrees} |
| moon_phase | {Phase of the Moon, 0 = 'new', 1 = 'full'} |
| waxing | {True if Moon's phase is increasing} |
| position_angle | {Position angle of Moon's bright limb, degrees} |
| moon_par_angle | {Parallactic angle of Moon, degrees} |
| | |
| lst_moon_rise | {Local sidereal time of moonrise, hours} |
| lst_moon_set | {Local sidereal time of moonset, hours} |
| gst_moon_rise | {Greenwich sidereal time of moonrise, hours} |
| gst_moon_set | {Greenwich sidereal time of moonset, hours} |
| lct_moon_rise | {Local civil time of moonrise, hours} |
| lct_moon_set | {Local civil time of moonset, hours} |
| ut_moon_rise | {Universal time of moonrise, hours} |
| ut_moon_set | {Universal time of moonset, hours} |
| azimuth_moon_rise | {Azimuth of moonrise, degrees} |
| azimuth_moon_set | {Azimuth of moonset, degrees} |
| moon_rs_valid | {True if Moon's rising and setting values are valid} |
| | |
| jd_new_moon | {Julian day number of new moon} |
| aml_new_moon | {Argument of Moon's latitude at new moon, degrees} |
| solar_eclipse | {True if a solar eclipse occurs} |
| jd_first_quarter | {Julian day number of first quarter} |
| jd_full_moon | {Julian day number of full moon} |
| aml_full_moon | {Argument of Moon's latitude at full moon, degrees} |
| lunar_eclipse | {True if a lunar eclipse occurs} |
| jd_last_quarter | {Julian day number of last quarter} |
| moon_phases_valid | {True if Moon phases are valid} |
| | |
| shadow_ra | {Right ascension of Earth's shadow, hours} |
| shadow_dec | {Declination of Earth's shadow, degrees} |
| shad_radec_valid | {True if Earth's shadow position is valid} |
| | |
| jd_first_contact | {Julian date of first contact in eclipse} |
| jd_mid_eclipse | {Julian date of mid eclipse} |
| jd_last_contact | {Julian date of last contact in eclipse} |
| jd_start_umbral | {Julian date of start of umbral phase} |
| jd_start_total | {Julian date of start of total phase} |

| | |
|---|---|
| jd_end_umbral | {Julian date of end of umbral phase} |
| jd_end_total | {Julian date of end of total phase} |
| rad_umbra | {Radius of Earth's umbra, degrees} |
| rad_penumbra | {Radius of Earth's penumbra, degrees} |
| magnitude | {Magnitude of eclipse } |
| annular | {True if the eclipse is annular} |
| | |
| epoch_daya | {Day number of epoch A, 1–31} |
| epoch_montha | {Month number of epoch A, 1–12} |
| epoch_yeara | {Year number of epoch A, BC negative} |
| epoch_dayb | {Day number of epoch B, 1–31} |
| epoch_monthb | {Month number of epoch B, 1–12} |
| epoch_yearb | {Year number of epoch B, BC negative} |
| epochs_valid | {True if the epochs are valid} |
| | |
| object_name | {Name of orbiting object} |
| inclination | {Inclination, degrees} |
| arg_perihelion | {Argument of the perihelion, degrees} |
| long_perihelion | {Longitude of the perihelion, degrees} |
| long_asc_node | {Longitude of the ascending node, degrees} |
| eccentricity | {Eccentricity of the orbit, 0–1} |
| mean_anomaly | {Mean anomaly at the epoch, degrees} |
| orbital_long | {Longitude at the epoch, degrees} |
| e_day | {Epoch day, 1–31} |
| e_month | {Epoch month, 1–12} |
| e_year | {Epoch year, BC negative} |
| orbital_period | {Period of the orbit, years} |
| daily_motion | {Daily motion in orbit, degrees} |
| semi_maj_axis | {Semi major axis length, au} |
| ell_elmt_valid | {True if elliptical orbital elements are valid} |
| | |
| perihelion_dist | {Perihelion distance, au} |
| par_elmt_valid | {True if parabolic orbital elements are valid} |
| | |
| osc_pos_valid | {True if osculating position is valid} |

# Bibliography

*Astronomical Algorithms*
by Jean Meeus (1991). Willmann-Bell Inc., PO Box 35025, Richmond, Virginia 23235, USA.

*The Astronomical Almanac*
Published annually jointly by US Government Printing Office, Washington, USA and Her Majesty's Stationery Office, London, UK.

*Astronomy with your Personal Computer*
by Peter Duffett-Smith (second edition 1990). Cambridge University Press, Shaftesbury Road, Cambridge, UK.

*Atlas of Historical Eclipse Maps: East Asia 1500 BC–AD 1900*
by F. R. Stephenson & M. A. Houlden (1986). Cambridge University Press, Shaftesbury Road, Cambridge, UK.

*ELP 2000-85: a semi-analytical lunar ephemeris adequate for historical times*
M. Chapront-Touze & J. Chapront (1988). *Astron. Astrophys.*, **190**, 342.

*Empirical transformations from UT to ET for the period 1800–1988*
L. D. Schmadel & G. Zech (1988). *Astronomische Nachrichten*, **309**, 219–21.

*Explanatory Supplement to the Astronomical Almanac*
edited by P. K. Seidelmann, US Naval Observatory (1992). University Science Books, 20 Edgehill Road, Mill Valley, California 94941, USA.

*Lunar Tables amd Programs from 4000 B.C. to A.D. 8000*
by M. Chapront-Touze & J. Chapront (1991). Willmann-Bell Inc., PO Box 35025, Richmond, Virginia 23235, USA.

*Planetary theories in rectangular and spherical variables. VSOP 87 solutions*
P. Bretagnon & G. Francou (1988). *Astron. Astrophys.*, **202**, 309–15.

*Practical Astronomy with your Calculator*
by Peter Duffett-Smith (third edition 1988). Cambridge University Press, Shaftesbury Road, Cambridge, UK.

# Glossary of astronomical terms

**aberration:** the apparent angular displacement of a celestial object from its geometric position, caused by the motion of the observer with respect to the object, and the finite speed of light.

**age of the Moon:** the angle between the Sun and the Moon measured at the Earth.

**altitude:** the angle up from the horizon.

**anomaly:** the angle at the focus (*true anomaly*) or the centre (*mean* and *eccentric anomalies*) of an orbital ellipse between the major axis and the orbiting body or its projection.

**aphelion:** the point in an orbit about the Sun which is most distant from the Sun.

**astrometric coordinates:** geocentric co-ordinates obtained after allowing for light time, but not for aberration or nutation. Such coordinates can be compared directly with the positions of stars given in catalogues.

**Astronomical Almanac:** a collection of tables predicting the positions and circumstances of astronomical phenomena. This title replaced both the *American Ephemeris* and *Nautical Almanac*, and the *Astronomical Ephemeris*, beginning with the 1981 edition.

**Astronomical Ephemeris:** see *Astronomical Almanac*.

**astronomical latitude:** the angle between the astronomical zenith and the equator.

**astronomical unit (AU):** (very slightly less than) the length of the semi-major axis of the Earth's orbit about the Sun.

**atmospheric refraction:** the apparent shift in the position of a celestial object caused by the bending of light rays by the atmosphere.

**azimuth:** the angle round from the north point measured on the horizon in the sense NESW.

**barycentre:** the centre of mass of an orbiting system. Hence the barycentre of the Solar System lies inside the Sun, but not at its centre because of the finite masses of all the planets in orbit about it.

**calendar:** system of accounting days in the year. The *Julian calendar*, introduced by Julius Caesar, divides the year into 365 days except for every fourth year (leap year) which has 366 days (leap year numbers are divisible exactly by 4). The *Gregorian calendar*, introduced by Pope Gregory in 1582, is the one in general use today. It reduces the discrepancy between the average length of the calendar year and the length of the tropical year by removing three days every 4 centuries; if the year ends in two zeroes it is only a leap year if it is divisible exactly by 400.

**celestial equator:** the curve of intersection of the extended plane of the Earth's equator with the celestial sphere.

**celestial sphere:** an imaginary sphere, usually centred on the Earth, of arbitrary large radius on the surface of which the stars can be considered to be fixed.

**circumpolar star:** a star whose angular distance from the north or south celestial pole is sufficiently small that it never dips below the horizon.

**comet:** a diffuse member of the Solar System, usually with a highly elongated elliptical or parabolic orbit, which becomes visible near the Sun. It has a bright head and a diffuse tail of variable length which always points away from the Sun.

**conjunction:** the moment when two celestial bodies occupy the same position in the sky, or share the same value of a coordinate when viewed from a particular place: thus *heliocentric conjunction*, and *conjunction in right ascension*.

**coordinate systems:** frames of reference by means of which the position of any point can be uniquely specified. In astronomy, the systems take their names from the fundamental planes on which they are based. Thus the *ecliptic coordinate system* measures longitude eastward from the first point of Aries in the plane of the ecliptic, and latitude northwards from it. The *equatorial coordinate system* measures right ascension eastward from the first point of Aries in the plane of the celestial equator, and declination northward from it. In the *horizon coordinate system*, the azimuth is measured eastward round the horizon from its north point, and altitude upwards from it (i.e. measured towards the zenith). The *galactic coordinate system* specifies position by longitude measured in the galactic plane round from the direction of the galactic centre as seen from the Sun, and by latitude measured perpendicular to the plane.

**coordinated universal time (UTC):** the time scale available from broadcast time signals. It differs from international atomic time *TAI* by a whole number of seconds, and is maintained within 0.9 s of *universal time* (strictly UT1) by the insertion of leap seconds as needed, usually at the ends of June and December.

**culmination:** the moment at which a celestial body crosses the observer's meridian. *Circumpolar stars* cross the meridian above the horizon twice in one day, giving *both upper culmination* and *lower culmination*.

**day:** the interval between two successive upper transits across the observer's meridian of the first point of Aries (sidereal day), of the Sun (solar day), or of a fictitious body moving at a uniform rate along the equator (mean solar day).

**declination:** in the equatorial coordinate system, the angle measured northward along the great circle through the object and the north celestial pole from the

equator to the object.

**dynamical time:** the family of time scales introduced in 1984 which replaced ephemeris time. See *time*.

**earthshine:** light reflected from the Earth which sometimes illuminates the dark portion of the Moon's disk, making it visible.

**eccentricity:** a measure of the degree of elongation of an ellipse, and equal to the ratio of the distance of a focus from the centre to the length of the semi-major axis. A circle has an eccentricity of 0, whilst the most-flattened ellipses have eccentricities approaching 1.

**eclipse:** the passage of the Moon through the Earth's shadow (*lunar eclipse*) or parts of the Earth through the Moon's shadow (*solar eclipse*). If, at the moment of greatest eclipse, the Moon or Sun is only partly obscured it is a *partial eclipse*; if completely obscured it is a *total eclipse*. If during a solar eclipse the Moon obscures the central part of the Sun's disk but leaves an unobscured ring around its edge, then it is an *annular eclipse*.

**ecliptic:** the plane containing the mean orbit of the Earth around the Sun.

**ellipse:** a type of regular closed curve, oval in shape, of which a circle is a special case. It is traced by a point moving in such a manner that it keeps constant the sum of its distances from two fixed points, each of which is called a focus of the ellipse. The longest diameter of the ellipse, which passes through both foci and the centre, is called the major axis, either half of which is called the semi-major axis. The axis through the centre perpendicular to the major axis is called the minor axis. (See also *eccentricity*.)

**ephemeris time (ET):** see *time*.

**epoch:** a particular moment specified as the reference point for which the given quantities are valid, or from which time is measured.

**equation of the centre:** a relation between the true anomaly, $v$, and the mean anomaly, $M$, which is an approximate solution of Kepler's equation. In its simplest form it is

$$v = M + 2e \sin M,$$

where $v$ and $M$ are expressed in radians, and $e$ is the eccentricity of the orbit.

**equation of the equinoxes:** apparent sidereal time minus mean sidereal time, taking into account the effect of nutation on the positions of the equinoxes.

**equation of time:** the difference between the mean solar time and the real solar time.

**equator:** the plane through the centre of the Earth, or other orbiting body, which is perpendicular to its spin axis.

**equinoxes:** the points where the plane of the ecliptic cuts the celestial equator, or the moments when the Sun is at those positions. The *vernal equinox*, also called the *first point of Aries*, is at the zero of right ascension and occurs at about 21st March each year. The *autumnal equinox* is at right ascension 12h, and occurs about 22nd September.

**extinction:** the colouring and attenuation of light as it travels through a medium; in particular, *atmospheric extinction*.

**figure of the Earth:** the true shape of the Earth. It is often approximated by a spheroid of revolution, a geometrical shape in which any cross-section parallel to the equator is a circle, whilst any cross-section through the north–south axis is an ellipse with the minor axis coincident with the line joining the north and south poles.

**first point of Aries:** the position on the celestial sphere of the vernal equinox.

**focus of an ellipse:** see *ellipse*.

**geocentric:** with respect to the centre of the Earth.

**geocentric coordinates:** coordinates measured with respect to the centre of the Earth. Hence the *geocentric latitude* is the angle between the *equator* and a point on the surface of the Earth, as measured at the centre.

**geocentric parallax:** the angle subtended at a heavenly body by the centre of the Earth and the point of observation on the Earth's surface.

**geostationary satellite:** a body orbiting the Earth in the plane of the equator in such a direction and at such a height that its orbital period equals one day, so that it keeps constant position with respect to the Earth's surface.

**gravity:** the mutual force of attraction between any two bodies which is proportional to the product of their masses and inversely proportional to the square of their separation.

**great circle:** any circle drawn on the surface of a sphere whose centre is coincident with the centre of the sphere.

**Greenwich mean time (GMT):** this is ambiguous and is not now used in *The Astronomical Almanac*. Its meaning in civil life is usually the same as *UTC*, though previously it has been used to mean *UT*. Before 1925 it was reckoned from Greenwich mean noon (12h UT).

**Greenwich meridian:** that half of the great circle on the surface of the Earth passing through the north and south poles and through the reference point in Greenwich, England. It is taken as the line of longitude 0°.

**heliocentric:** measured at the centre of the Sun.

**horizon:** the line of intersection between the sky and the Earth. For astronomical purposes, an observer's horizon is taken to be the great circle on the celestial sphere on which every point is 90 degrees distant from the zenith point.

**horizontal parallax:** the geocentric parallax when the celestial body is on the observer's horizon; hence *equatorial horizontal parallax* when the observer is also on the equator.

**hour angle:** the difference between the local sidereal time and the right ascension. It is a measure of the sidereal time since the object crossed the local meridian at upper transit.

**inclination of orbit:** the angle between the plane of the orbit and the plane of the ecliptic.

**inner planet:** a planet with an orbit whose semi-major axis is less than that of the Earth; that is, the planets Mercury and Venus.

**international atomic time:** see *time*.

**Julian day number:** the number of Julian days (including the fraction of a day) that have elapsed since the epoch Greenwich mean noon of January 1st 4713 BC. For midday on January 1st 2000 (i.e. 2000 Jan 1.5) its value is 2451545.0.

**Kepler's equation:** the relation between the mean anomaly, $M$, the eccentric anomaly, $E$, and the eccentricity, $e$, of the form

$$E - e \sin E = M,$$

where the angles are expressed in radians. (See also *equation of the centre*.)

**latitude:** the coordinate expressing the angle (north positive, south negative) measured along a great circle perpendicular to a fundamental plane, hence *ecliptic latitude* and *galactic latitude*. On the Earth, the *geographical latitude* is measured with respect to the equator. The ecliptic latitude can either be measured at the earth (geocentric) or at the sun (heliocentric).

**light time:** the time it takes the light from a celestial body to reach an observer.

**longitude:** the coordinate expressing the angle round from a fixed direction measured in a fundamental plane, hence *ecliptic longitude* and *galactic longitude*. On the Earth, the *geographical longitude* is measured along the *equator*. The ecliptic longitude can either be measured at the Earth (geocentric) or at the Sun (heliocentric).

**lower culmination:** the moment when a celestial body crosses the observer's meridian at its lowest point, usually below the horizon unless the object is circumpolar.

**luni-solar precession:** the slow retrograde motion of the first point of Aries along the equator caused by the combined effects of the gravity of the Sun and the Moon on the slightly non-spherical Earth.

**magnitude:** (i) the unit defined on a logarithmic scale which measures the visual brightness of a celestial object considered as a point. Increasing magnitudes denote dimmer objects, and the dimmest object visible with the naked eye has a magnitude of about 6.

(ii) in a lunar eclipse, the fraction of the lunar diameter obscured by the shadow of the Earth at the moment of greatest eclipse, measured along the common diameter.

(iii) in a solar eclipse, the fraction of the solar diameter obscured by the Moon at the moment of greatest eclipse, measured along the common diameter.

**mean Sun:** a fictitious heavenly body which moves at a uniform rate along the celestial equator, making one complete revolution in the same time (one year) as the real Sun takes to complete its apparent orbit about the Earth.

**meridian:** that half of a great circle which is terminated at the north and south poles. On the Earth a meridian is a line of longitude. On the celestial sphere, the meridian which passes through the zenith is called the observer's meridian.

**modified Julian date:** the number of Julian days elapsed since 1858 November 17.0.

**month:** the period taken by the Moon to make one complete circuit of its orbit from reference point to reference point. The *draconic* or *nodal month* takes the ascending node as the reference and is equal to 27.2122 mean solar days. The *sidereal month* is reckoned against the background of stars and is equal to 27.3217 mean solar days. The Sun is used as the reference point for the *synodic month* of 29.5306 mean solar days, the perigee in the *anomalistic month* of 27.5546 mean solar days.

**nadir:** the point on the celestial sphere which is diametrically opposite the zenith.

**node:** the point on the celestial sphere where the great circle representing the orbit cuts the great circle representing the plane of the ecliptic. The point where the orbiting body is moving from south to north (i.e. from below the ecliptic to above it) is called the ascending node; the other is the descending node.

**noon:** the instant at which the Sun (mean or real) crosses the observer's meridian at upper transit.

**north celestial pole:** the point at which the projection of the Earth's rotation axis through the north pole cuts the celestial sphere.

**nutation:** a small periodic wobbling motion of the Earth's rotation axis.

**obliquity of the ecliptic:** the angle at which the plane of the ecliptic is inclined to the plane of the equator.

**observer's meridian:** see *meridian*.

**opposition:** the moment when two celestial bodies occupy opposite positions in the sky, or have longitudes different by 180°, when viewed at a particular place.

**orbit:** the path through space followed by a body gravitationally attracted to another body.

**orbital elements:** the quantities that need to be known in order to specify an orbit uniquely.

**osculating elements:** the elements describing the orbit followed by a body at a particular point if all perturbing influences vanish. Since perturbations disturb the true orbit of any member of the Solar System, the osculating elements are constantly changing.

**outer planets:** those planets with orbits having semi-major axes larger than that of the Earth. The major outer planets are Mars, Jupiter, Saturn, Uranus, Neptune, and Pluto.

**parabolic orbit:** an orbit in which the velocity at any point is equal to the escape velocity.

**parallax:** the amount by which the apparent position of a celestial object shifts as the point of observation is changed.

**penumbra:** the outer portion of a shadow where the light is only partially cut off.

**perigee:** the point on an orbit about the Earth which is nearest the Earth.

**perihelion:** the point of closest approach to the Sun in an orbit about the Sun.

**period of orbit:** the time taken by an orbiting body to make one complete circuit.

**perturbations:** deviations from true parabolic or elliptical motion caused by the gravitational fields of other members of the Solar System.

**phase:** (i) *of Moon or planet*: the fraction of the area of the disk which is illuminated. When the dark side of the Moon

faces the Earth, the phase is zero and it is *new moon*. At the *first quarter* and *third quarter*, the phase is equal to one half and the Moon is in quadrature. *Full moon* has a phase of one. Whenever the phase of the Moon is greater than one half, the Moon is described as gibbous.

(ii) *of an eclipse*: the stage of a lunar or solar eclipse during which the eclipsed body is partly obscured (*partial phase*) or totally obscured (*total phase*). During a lunar eclipse, the Moon is in the penumbra of the Earth's shadow during the *penumbral phase*, and is partially or totally in the umbra during the *umbral phase*. The partial and total phases occur during the umbral phase.

**planet:** a solid body in a closed orbit about a star. In our own Solar System, the major planets are (in order of increasing distance from the Sun) Mercury, Venus, Earth, Mars, Jupiter, Saturn, Uranus, Neptune, and Pluto.

**polar distance:** the angle on the celestial sphere from the celestial pole.

**pole:** the point on a sphere where a line drawn from it to the centre of the sphere is perpendicular to a given plane. Hence *pole of the ecliptic* and *pole of the equator* (each has two poles called north and south).

**position-angle:** a celestial angle measured 0° to 360° eastwards from the north.

**precession:** see *luni-solar precession*.

**prograde motion:** motion in the same sense as that of all the planets about the Sun in the Solar System. When looking down on the Solar System from the north celestial pole, prograde motion is counter clockwise.

**proper motion:** the apparent angular motion per year of a star due to its motion in space relative to the Solar System.

**radius vector:** the line joining the principal focus to the position of the orbiting body on its orbital ellipse.

**refraction:** see *atmospheric refraction*.

**retrograde motion:** motion in the opposite sense to that of all the planets about the Sun in our Solar System. When looking down at the Solar System from the north celestial pole, retrograde motion is clock-wise.

**right ascension:** in the equatorial coordinate system the angle measured eastward from the first point of Aries in the plane of the equator.

**rising:** the moment when a celestial body crosses the horizon on the way up.

**Saros cycle:** the period of 18 years 11 days and 8 hours after which the pattern of lunar and solar eclipses tends to repeat.

**second (SI second):** the unit of time for the *TAI* scale defined to be exactly 9 192 631 770 cycles of radiation corresponding to the transition between two hyperfine levels in the ground state of caesium 133.

**semi-major axis:** see *ellipse*.

**setting:** the moment when a celestial body crosses the horizon on the way down.

**solar elongation:** the angle between the lines of sight to the Sun and to the celestial body in question.

**Solar System:** the Sun and all the bodies, planets, comets, and asteroids in closed orbits about it.

**solstice:** the points at which the apparent celestial longitude of the Sun is 90° or 270°, or the moments when the Sun is at either of these points. They occur around June 21st and December 21st.

**synodic period:** the time between successive conjunctions in longitude.

**terminator:** the line marking the boundary between the dark and sunlit hemispheres of a member of a solar system.

**terrestrial dynamical time (TDT):** see *time*.

**time:** (i) *atomic time:* time measured with respect to the natural period of oscillation of an atomic system. Caesium beam clocks currently constitute the most precise time-keepers available, and the SI unit (second) of atomic time is defined in terms of the caesium 133 atom. *International atomic time (TAI)* is the continuous scale resulting from analyses by the Bureau International des Poids et Mesures of atomic time standards in several countries, starting from the epoch 1958 January 1st. *Coordinated universal time (UTC)* is the time scale distributed by standard time services and is tied to both TAI and universal time (UT) in such a manner that (a) it differs from TAI by a whole number of seconds, and (b) it is never more than 0.9 s different from UT (strictly UT1). This is achieved by the introduction of leap-seconds into UTC from time to time. UTC constitutes the basis for legal time-keeping in most parts of the world. *Terrestrial dynamical time (TDT)* is used as the argument in theories of celestial dynamics and in the compilation of the *Astronomical Almanac*. It is equal to TAI + 32.184 s. It replaced *ephemeris time (ET)* in 1984, which was itself derived from analyses of the Moon's motion. *Barycentric dynamical time (TDB)* is the time argument for equations of motion with respect to the barycentre of the Solar System. It differs from TDT by small variable offsets which take account of the influence of gravity on time, never amounting to more than about 0.002 s.

(ii) *solar time:* time measured with respect to the motion of the Sun or a fictitious body, no longer used, called the mean Sun (hence *mean solar time*). *Universal time* is, broadly speaking, the mean solar time as measured on the Greenwich meridian. It is formally defined by a mathematical formula as a function of *sidereal time* (see below), and is thus determined from observations of the stars. A direct application of the formula gives UT0; with a small correction for polar motion the scale UT1 is obtained. (The term UT usually implies UT1). *British summer time (BST)* is one hour ahead of UT and is an example of *daylight saving time* in which the time is adjusted to make the working day fit more conveniently into the daylight hours.

(iii) *sidereal time:* time measured with respect to the apparent motion of the stars. The *local sidereal time* at any place is equal to the hour angle of the first point of Aries; local sidereal time on the Greenwich meridian is called Greenwich sidereal time. The difference between apparent and mean sidereal times is called the equation of the equinoxes, and takes

**account** of nutation. It may be as much as 1.2 seconds.

**time zone:** a longitudinal strip on the surface of the Earth in which the *zone time*, a whole number of hours before or after UT, is adopted as the local civil time by national or international agreement.

**topocentric:** as measured at the observer's location on the surface of the Earth.

**transit:** the moment at which a celestial body crosses the observer's meridian.

**twilight:** that period of semi-darkness after sunset or before sunrise during which the Sun's zenith distance (measured at its centre) is between 90° 50′ and 96° *(civil twilight)*, 96° and 102° (nautical twilight), and between 102° and 108° (astronomical twilight).

**umbra:** the inner portion of a shadow where the light is completely obscured.

**universal time (UT):** see *time*.

**upper culmination:** the moment when a celestial body crosses the observer's meridian at its highest point.

**vernal equinox:** see *equinox*.

**year:** the interval between two successive passages of the Sun through a reference point. A particular point amongst the stars is used as the reference in the *sidereal year*, equal to 365.2564 mean solar days. The *tropical year*, 365.242 191 mean solar days, uses the first point of Aries as its reference. When no qualifying adjective is used with the word 'year', it is usually the tropical year that is meant. Perturbations to the Earth's orbit by the other planets cause small changes in the Earth's orbital elements. The *anomalistic year*, 365.2569 mean solar days, is the interval between two successive passages of the Sun through perigee. The *Besselian year*, not used since 1984, is the period of one complete revolution in right ascension of the fictitious mean sun as defined by Newcomb. It is almost the same as the tropical year, but begins when the right ascension of the Sun is exactly 240°, which falls very near the beginning of the civil year.

**zenith:** the point directly overhead at the observer. The *zenith angle* or *zenith distance* of a star is the angle between the star and the zenith.

**zone correction:** the number of hours that needs to be added to UT to get the *zone time* (see *time zone*).

# Index